OXFORD WORLD'S **P9-BJE-711**

THE OXFORD SHAKESPEARE

General Editor · Stanley Wells

The Oxford Shakespeare offers new and authoritative editions of Shakespeare's plays in which the early printings have been scrupulously re-examined and interpreted. An introductory essay provides all relevant background information together with an appraisal of critical views and of the play's effects in performance. The detailed commentaries pay particular attention to language and staging. Reprints of sources, music for songs, genealogical tables, maps, etc. are included where necessary; many of the volumes are illustrated, and all contain an index.

JAY L. HALIO, the editor of *The Merchant of Venice* in the Oxford Shakespeare, is Professor of English at the University of Delaware.

THE OXFORD SHAKESPEARE

Currently available in paperback

The rest of the plays are forthcoming

OXFORD WORLD'S CLASSICS

WILLIAM SHAKESPEARE

The Merchant of Venice

Edited by
JAY L. HALIO

OXFORD
UNIVERSITY PRESS

OXFORD
UNIVERSITY PRESS

Great Clarendon Street, Oxford OX2 6DP

Oxford University Press is a department of the University of Oxford.
It furthers the University's objective of excellence in research, scholarship,
and education by publishing worldwide in

Oxford New York

Athens Auckland Bangkok Bogotá Buenos Aires Cape Town
Chennai Dar es Salaam Delhi Florence Hong Kong Istanbul Karachi
Kolkata Kuala Lumpur Madrid Melbourne Mexico City Mumbai Nairobi
Paris São Paulo Shanghai Singapore Taipei Tokyo Toronto Warsaw

with associated companies in Berlin Ibadan

Oxford is a registered trade mark of Oxford University Press
in the UK and in certain other countries

Published in the United States
by Oxford University Press Inc., New York

British Library Cataloguing in Publication Data

Data available

Library of Congress Cataloging in Publication Data

Data available

ISBN 0–19–812925–4 (hbk.)
ISBN 0–19–283424–X (pbk.)

8

Printed in Great Britain by
Clays Ltd, St Ives plc

PREFACE

I AM deeply grateful to many who have assisted in the work of this edition, especially Thomas Clayton, who read all of the Introduction, to the General Editor, Stanley Wells, who made many useful suggestions and corrections, and to Edwin F. Pritchard, who was a most careful copy-editor. Had it not been for their painstaking labour on my behalf, many more errors and faults would have escaped my notice; those that remain are entirely my responsibility. George Walton Williams and Richard Kennedy not only kindly sent me unpublished manuscripts that illuminated aspects of the text, but also read an earlier version of the textual analysis. James Shapiro made several valuable comments on 'Shakespeare and Semitism'. Meghan Cronin and Christine Volonte, graduate assistants in English at the University of Delaware, helped in the research and checking the typescript. Marcia Halio also read the Introduction in manuscript, and Frances Whistler assisted greatly with the proof-reading. Finally, I owe a large debt of thanks to the personnel and resources of the Morris Library of the University of Delaware and to those of the Folger Shakespeare Library and the Shakespeare Centre Library, Stratford-upon-Avon, for helping to bring this edition to completion.

Note: As this edition was being printed, John Gross's valuable study, *Shylock: Four Hundred Years in the Life of a Legend*, appeared. Unfortunately, I was unable to make use of its many cogent arguments or information, either in the Introduction or the Commentary. Gross is especially useful in some aspects of the stage history, since he surveys a great many more productions than space permits here. He is also extremely interesting in tracing *The Merchant*'s role, and specifically Shylock's, in the history of anti-Semitism in the western world, a major focus of his book, as its subtitle indicates. In his analysis of the background to the play, or to Semitism in England before and during the time of Shakespeare, he covers much the same ground as I do in the first part of this Introduction, but he does not add anything substantially new or different. For that, I suspect we must await the further researches of James Shapiro and others who are investigating the attitudes and actions towards Jews in England and on the Continent during the Middle Ages and the Renaissance. Meanwhile, we have Gross's survey not only of the stage history, but also of the critical commentary down through the centuries, including an illuminating chapter on the responses and reactions to the play by Jewish writers, such as Heinrich Heine, Italo Svevo, Marcel Proust, and Ludwig Lewisohn.

CONTENTS

LIST OF ILLUSTRATIONS

GENERAL INTRODUCTION

ANY approach to understanding Shakespeare's *The Merchant of Venice* inevitably includes a discussion of the vexed question of its alleged anti-Semitism. This Introduction to the play therefore confronts the question directly, focusing on the background against which the play must be considered and a comparison with another play famous, or infamous, for its portrayal of a Jew, Marlowe's *The Jew of Malta*. From thence a discussion of the *Merchant*'s more immediate sources and its date continues, followed by a detailed analysis of the play itself, which emphasizes its ambiguities, inconsistencies, and internal contradictions. This discussion naturally leads into a survey of the play's performance history, particularly its representation of the dominant character, Shylock, and the major ways he has been portrayed. The Introduction concludes with a discussion of the text and the editorial procedures followed in this edition.

Shakespeare and Semitism

Shakespeare's attitude toward Jews, specifically in *The Merchant of Venice*, has been the cause of unending controversy. Recognizing the problem, in the Stratford-upon-Avon season of 1987 the Royal Shakespeare Company performed *The Merchant of Venice* back-to-back with a production of Marlowe's *The Jew of Malta*.[1] *The Jew of Malta*, played as a very broad heroic comedy, was evidently intended to contrast with Shakespeare's play and disarm criticism, such as the RSC had experienced earlier, in 1983, with a less successful production of *The Merchant*. To reinforce the new strategy, Antony Sher, a South African Jew, was cast as Shylock.[2] It almost worked, but not quite. Sher was largely a sympathetic Shylock, with swastikas and other anti-Semitic slurs used to underscore the

---

[1] *The Merchant* was performed on the main stage, with *The Jew* at the Swan Theatre.

[2] Sher was not the first Jewish actor recently to essay the role on the British stage. David Suchet, for example, had played Shylock at the RSC in 1981.

money-lender as victim; however, the trial scene portrayed Shylock as extremely bloodthirsty. Interpolating some extra-Shakespearian stage business, borrowed from the Passover Haggadah, the RSC and Sher indicated that cutting Antonio's pound of flesh was tantamount to a religious ritual of human sacrifice. Of course, nothing could be further from Jewish religious practice or principles, the aborted sacrifice of Isaac in Genesis 22 being the archetype of Jewish opposition to human sacrifice.[1] In the event, Antony Sher's Shylock was not far removed from Alun Armstrong's Barabas.

Looking closely at both Marlowe's play and Shakespeare's will clarify the attitudes towards Semitism in those dramas, but the background against which both were conceived is also important. Jews had been officially banished from England since the Expulsion of 1290 by King Edward I, but the eviction was not quite so thoroughgoing as was hoped. A few Jews, whether converts or not, remained in England in the intervening period before Oliver Cromwell invited them to return in the seventeenth century. There is sufficient evidence for this assertion, but whether Shakespeare or Marlowe actually knew any Jews may be irrelevant.[2] In their plays they wrote not from personal experience but from a tradition that had evolved both in England and on the Continent of the Jew as alien, usurer, member of a *race maudite*.[3] In Marlowe's case, the tradition of the amoral machiavel was even more important than that of the money-lender.

In these post-Holocaust days, it may be difficult for us to conceive how Jews were regarded and treated in Europe, including England, during the Middle Ages. They had few rights and could not claim inalienable citizenship in any country. Typically, they depended upon rulers of the realm

[1] Cf. Hermann Sinsheimer, *Shylock: The History of a Character* (1947; repr. New York, 1964), 133–4.

[2] See Danson, 60. A Jewish merchant from Venice, Alonzo Nuñez de Herrera (Abraham Cohen de Herrera), was captured in Essex's raid on Cadiz and brought to London as one of forty hostages in 1596, where he remained until 1600. It is unlikely, however, that he bears any resemblance to Shakespeare's Shylock or (though born in Florence) to Marlowe's Barabas. See Richard H. Popkin, 'A Jewish Merchant of Venice', *SQ* 40 (1989), 329–31.

[3] See Leo Kirschbaum, 'Shylock and the City of God', *Character and Characterization in Shakespeare* (Detroit, 1962), 7–8.

for protection and such rights as they might enjoy. In the thirteenth century in England, for example, under Henry III and Edward I, they were tantamount to the king's chattel. The king could—and did—dispose of them and their possessions entirely as he chose. Heavy talladges, or taxes, were imposed upon Jews—individually and collectively—to support the sovereign's financial needs, and when the moneys were not forthcoming, imprisonment and/or confiscation usually followed. At the same time, the Church vigorously opposed the existence of Jews in the country, but as they were under the king's protection the Church was powerless to do more than excite popular feeling against them.

Contrary to common belief, not all Jews were money-lenders, although usury was one of the few means to accumulate such wealth as they had. Many Jews were poor and served in humble, even menial capacities.[1] But as non-believers in Christ, they were a despised people, however useful, financially and otherwise (as doctors, for instance). Near the end of the thirteenth century, when Edward had practically bankrupted his Jews, who found it impossible to meet his increasingly exorbitant demands for payments, the king decided to play his last card—expulsion. This act was not only satisfying to the Church, but it provided the king with the last bit of income from that once profitable source. Since everything the Jews owned belonged to the king, including the debts owed them as money-lenders or pawnbrokers, the king became the beneficiary of those debts as well as everything else of value. Although Edward relieved the debtors of the interest on their loans and made some other concessions, he hoped to realize a sizeable amount of money eventually, however much he might later regret the termination—forever?—of this once lucrative source of income.[2]

Doubtless, some Jews preferred conversion to expulsion in England, as later in Spain under the Inquisition, and they took shelter in the *Domus conversorum*, the House of Converts. This institution dates from the early thirteenth century and

---

[1] Cecil Roth, 'A Day in the Life of a Medieval English Jew', *Essays and Portraits in Anglo-Jewish History* (Philadelphia, 1962), 36.

[2] See H. G. Richardson, 'The Expulsion', *The English Jewry under Angevin Kings* (1960), 213–33.

was an effort by the Dominicans, assisted by the king, to convert Jews to Christianity. The *Domus conversorum* in what is now Chancery Lane in London lasted well into the eighteenth century. Although at times few if any converts of Jewish birth lived there, in the centuries following 1290 it sheltered several from Exeter, Oxford, Woodstock, Northampton, Bury St Edmunds, Norwich, Bristol, as well as London and elsewhere where Jews had lived before being expelled.[1] After the Expulsion, some Jews entered the realm for one reason or another, either as travellers and merchants, as refugees from Spain and Portugal, or as invited professionals, such as the physicians who treated Henry IV in his illness and the engineer, Joachim Gaunse, who helped found the mining industry in Wales in the sixteenth century.[2] Small settlements of Marranos, or crypto-Jews, can be traced in London and Bristol during the reigns of Henry VIII and Elizabeth I.[3]

But if Jews were scarce and those few who remained were forbidden to practise their religion openly, they were not forgotten, either in history or legend, and certainly not in the popular imagination, as ballads and other literature indicate.[4] As mystery plays grew and flourished, Old and New Testament stories were dramatized, with Jews occupying a prominent place in both. One recent scholar has suggested that the contrast between the biblical Jews of the Old Testament and those of the New Testament, particularly in plays dealing with the Crucifixion and events leading up to it, resulted in a 'dual image' of the Jew. On the one hand, 'he excites horror, fear, hatred; but he also excites wonder, awe, and love'.[5] The

[1] A. M. Hyamson, *A History of the Jews in England* (1908), 125–33.

[2] See Cecil Roth, *A History of the Jews in England*, 3rd edn. (Oxford, 1964), 132–48; E. N. Calisch, *The Jew in English Literature* (1909; repr. Port Washington, NY, 1969), 41–2; Hyamson, *History of the Jews in England*, 135–6.

[3] Lucien Wolf, 'Jews in Tudor England', in C. Roth (ed.), *Essays in Jewish History* (1934), 73–90; 'Jews in Elizabethan England', *Transactions of the Jewish Historical Society*, 11 (1924–27), 1–33; C. J. Sisson, 'A Colony of Jews in Shakespeare's London', *Essays and Studies*, 23 (1937), 38–51. Far from being oppressed, the Marranos in Shakespeare's London, Sisson says, reaped the rewards of compromise and submission to law, carrying on trade or entering professions, so long as they did not flaunt their real nonconformity.

[4] See Calisch, *The Jew in English Literature*, 51–4, and Warren D. Smith, 'Shakespeare's Shylock', *SQ* 15 (1964), 193–4.

[5] Harold Fisch, *The Dual Image: The Figure of the Jew in English and American*

4

examples of Judas and the Pharisees in the Corpus Christi plays must have supported common belief in the Jew as an incarnation of the devil;[1] on the other hand, the patriarchs, Moses, Daniel, the prophets, and other figures appear as heroes, symbols or presentiments of patience, constancy, and other Christian virtues.[2] Moreover, Christian theology, as represented in the epistles of St Paul, as in Romans 11 for example, argues for the redemption of Israel through conversion to Christianity. The Jews of post-biblical history, therefore, must be present not only 'as witness to the final consummation of the Christian promise of salvation', but as a participant in it.[3] If Jews were shunned as a pariah race, they also had to be preserved for the ultimate Christian fulfilment; hence, the 'dual image', and the dialectic of Christian thought and feeling regarding them.

The significance of this twofold attitude, and of historical actions against Jews in England and elsewhere, is apparent in *The Merchant of Venice*. Earlier, it appears in such works as the Croxton *Play of the Sacrament*, written in the latter half of the fifteenth century. Its principal character is Aristorius, a Sicilian merchant with connections all over the known world, from Antioch to Holland, and from the Brabant to Turkey. Jonathas, a wealthy Jew, approaches him intent on testing the efficacy of the Holy Sacrament, in which he utterly disbelieves. Only the riches he has acquired—gold and precious gems—mean anything to him. Jonathas bribes Aristorius with a hundred pounds to steal the holy wafer from the church and give it to him, whereupon miraculous events occur. When he and his four compatriots strike the Host with their daggers (a re-enactment of the Crucifixion), it begins to bleed. Jonathas picks it up and tries to put it in a cauldron of boiling oil, but it sticks to his hand and he is unable to get free of it. In the succeeding comic turmoil, Jonathas loses his hand; the water in the cauldron turns to blood after the

Literature (New York, 1971), 13. Cf. Calisch, *The Jew in English Literature*, 54–6.

[1] Cf. *Merchant* 2.2.25, where Lancelot Gobbo says, 'Certainly the Jew is the very devil incarnation'.

[2] Fisch, *Dual Image*, 16–18.

[3] Ibid. 14. Compare Danson, who cites Fisch, *Dual Image*, 165–9.

Host and his hand are thrown into it; and when the Host is finally removed and thrown into a hot oven, the oven bursts, bleeding from its cracks, and an image of the crucified Christ emerges. A dialogue, in English and Latin, ensues between the image, Jonathas, and the others, in which Jesus sorrowfully asks why they torment him still and refuse to believe in what he has taught:

> Why blaspheme yow me? Why do ye thus?
> Why put yow me to a newe tormentry?
> And I dyed for yow on the crosse![1]

The Jews are contrite and repent, converting to Christianity; whereupon Jonathas's hand is restored and Aristorius, abjectly penitent, is absolved from his sin.

The representation of the Jew in this fifteenth-century miracle play combines the attributes of physical mutilation (blood sacrifice) and commercial malpractice, as Edgar Rosenberg remarks.[2] But beneath its obviously broad comedy, it also shows a strong impulse on the part of the unknown playwright to encourage regeneration through conversion. Later, in Robert Wilson's play, *The Three Ladies of London* (1584), the Jew Gerontus appears as the hero and Mercadore, an Italian merchant, is the comic villain who speaks in broken English and is willing to embrace Islam rather than pay Gerontus the debt he owes him. In the event, Gerontus prefers to surrender the debt obligation so that Mercadore will not be driven to apostasy, but even so Mercadore is unrepentant, and both are finally brought before an upright judge, who passes appropriate sentence.

Generous Gerontus, however, is hardly typical of the Jewish stereotype in Elizabethan literature. The scoundrels Zadoch and Zachary in Thomas Nashe's *The Unfortunate Traveller* (1594) are much more like what we might expect, or the Jewish apothecary who poisons Bajazeth and Aga in *The Tragical Reign of Selimus* (1594). These and other comic vil-

---

[1] *The Play of the Sacrament*, in *Chief Pre-Shakespearean Dramas*, ed. Joseph Quincy Adams (Cambridge, Mass., 1924), 257, ll. 651–3.

[2] 'The Jew in Western Drama: An Essay and A Checklist', *Bulletin of the New York Public Library*, 72 (1968), 442–91; repr. in Edward Coleman, *The Jew in English Drama: An Annotated Bibliography* (New York, 1970), 1–50. The reference is to the reprint, p. 7.

lains may owe something to the notoriety caused by the trial and execution in 1594 of Roderigo Lopez, the Portuguese convert, who had been Queen Elizabeth's physician.[1] They may also owe a good deal to Marlowe's Barabas, the protagonist of *The Jew of Malta* (c.1589), a direct descendant of the vice figure in the morality plays[2] as well as the villainous Jews of the mystery and miracle plays. By a stroke of genius, Marlowe combined these elements with the popular conception of the Italian machiavel to produce the comic and heroic villainy of Barabas, a major dramatic figure and an extraordinarily powerful one.

While all three aspects of his character are important, Barabas as a comic machiavel emerges as the dominant one, suggested from the start when Barabas's opening soliloquy immediately follows Machiavel's prologue. Usury, so often associated with Jews, is not nearly as significant as Barabas's evident delight in his multifarious scheming. Owing loyalty to no one—not even, finally, his daughter Abigail—he proceeds despite setbacks to confound his enemies, until he ultimately and comically overreaches himself; or rather, until his enemies, Ferneze, the governor of Malta, and his knights surpass his treachery. For neither the Christians nor the Jews nor the Turks who threaten Malta emerge as the moral centre of this play, which instead substitutes wit and the ability to implement 'policy' as the controlling force. 'Marlowe is not finally interested, as Shakespeare is', Rosenberg says, 'in questions which touch deeply on the nature of justice, is even less interested in legalistic quibbles; he enjoys the spectacle of these depraved noblemen of passion trying to cut each other's throats' (pp. 20–1).

But what of Barabas's Jewishness and its role in the drama? As an alien figure, an outsider, the Jew might be associated with the amoral machiavel, except that Jews, as representative of the Old Testament, had a strict moral code of their own. In his references to the patriarchs and biblical story, Barabas

[1] See Hyamson, *History of the Jews in England*, 136–40, for an account of this episode, and compare Sinsheimer, *Shylock*, 62–8, who notes the crowd's derision as Lopez was executed.

[2] See David Bevington, *From 'Mankind' to Marlowe* (Cambridge, Mass., 1962), 218–33.

confirms his Jewish heritage, but in the process he comically perverts it. For example, he equates the riches he has acquired with the blessings promised to the Jews (1.1.101–4).[1] When threatened with a tax needed to pay the Turkish tribute—a tax reminiscent of Edward I's 'talladges'—Barabas does not seek refuge in conversion to Christianity; but his hesitation results in confiscation of his wealth. Only his craftiness in hiding the better part of his fortune prevents complete destitution. His revenge later is to have the governor's son killed in a duel with his daughter's rival suitor, Mathias—the start of a series of murders and atrocities accomplished through duplicity and deception that characterize the hero-villain.

Duplicity and deception provide the link between Barabas's Jewishness and Machiavellism, at least in the popular imagination to which Marlowe appealed. Barabas implies the connection in his soliloquy brooding upon Ferneze's unjust confiscation of his property:

> I am not of the tribe of Levi, I,
> That can so soon forget an injury.
> We Jews can fawn like spaniels when we please,
> And when we grin, we bite; yet are our looks
> As innocent and harmless as a lamb's.
> I learned in Florence how to kiss my hand,
> Heave up my shoulders when they call me dog,
> And duck as low as any barefoot friar,
> Hoping to see them starve upon a stall. . . .
> (2.3.18–26)

Florence, of course, was Machiavelli's city. Unlike Venice, it was not known for harbouring many Jews. The soliloquy occurs shortly before Barabas purchases the slave Ithamore, a Turk who rivals him in treachery, especially directed against Christians (see 2.3.171–212). Since Marlowe includes Christian treachery as well in his play, most prominently at the end, it is clear that he enjoys attacking hypocritical professors of all three major religions, not solely the Jewish machiavel.

---

[1] References are to *The Complete Plays of Christopher Marlowe*, ed. I. Ribner (New York, 1963). For the Deity's promises to Abram, see Gen. 12: 1–3, 7; 15: 5; 17: 4–8, 16.

Shakespeare also attacks Christian hypocrisy, as modern commentators have frequently noted, specifically in Shylock's speech on Christian slave-holding (4.1.89–99).[1] But the conception of Shylock is altogether different from Marlowe's Barabas, notwithstanding the fact that both authors drew upon the same historical and literary backgrounds. Whereas Marlowe seems intent on a virtuoso display of comic villainy, with little regard for serious or deep character motives after Acts 1 and 2, Shakespeare concentrates upon Shylock's complex nature and the relationships of justice and mercy that lie at the heart of his play. If Shylock is another version of the villainous Jewish money-lender, and like Barabas a comic villain, he is also something more—the first stage Jew in English drama who is multi-dimensional and thus made to appear human.

Scholars, including myself, have looked elsewhere in Shakespeare's work for references to Jews and from them to discover more about his attitude. The references, such as those in *The Two Gentlemen of Verona*, *1 Henry IV*, and *Macbeth*, are hardly complimentary, though usually offhandedly remarked and consistent with the dramatic character. The references in *Love's Labour's Lost*, *A Midsummer Night's Dream*, and *Much Ado About Nothing*, on the other hand, are clearly humorous ones, depending in part on word-play to gain effect.[2] No allusions at all appear in the poems or sonnets. Anti-Semitic slurs thus do not appear to be important in Shakespeare's vocabulary or his thinking, with the outstanding exception of *The Merchant of Venice*. There, in the view of some critics, Shakespeare unleashed a venomous attack upon Jews—not only money-lenders and usurers, but all Jews.[3] To cite only one piece of evidence, Shylock is rarely referred to by name; instead, he is typically referred to or addressed as 'Jew', a

[1] See e.g. John R. Cooper, 'Shylock's Humanity', *SQ* 21 (1970), 122.

[2] M. J. Landa, *The Jew in Drama* (1926; repr. Port Washington, NY, 1968), 70–1. N. Nathan, 'Three Notes on *The Merchant of Venice*', *Shakespeare Association Bulletin*, 23 (1948), 158, 161–2 n. 9, adds seven references to 'Jewry', but says none are abusive. But allusions to the treachery of Judas, as in *As You Like It* 3.4.7–11, and *Richard II* 4.1.170, must also be included in any complete list.

[3] See e.g. D. M. Cohen, 'The Jew and Shylock', *SQ* 31 (1980), 53–63, esp. 54–5; also Nathan, 'Three Notes', 157–60, and Hyam Maccoby, 'The Figure of Shylock', *Midstream*, 16 (Feb. 1970), 56–69.

term then as now (in some quarters) of considerable contempt.[1]

Despite this fact, or rather in addition to it, complicating Shakespeare's attitude and our understanding of it, are other aspects of Shylock's character. These have enabled some actors, notably Henry Irving and Laurence Olivier, to portray Shylock as sympathetic, someone more sinned against than sinning, in short a tragic figure. A certain amount of textual adaptation, such as cutting Shylock's long aside, 'How like a fawning publican he looks. | I hate him for he is a Christian', etc. (1.3.38–49), is of course essential for this interpretation, although the dramatist otherwise endowed his comic villain with sufficient depth to permit the tragic emphasis. But it needs to be stressed that despite Shylock's depths, his very human traits,[2] Shakespeare's initial conception of him was essentially as a comic villain, most likely adorned with a red wig and beard and a bottle nose, but not a middle-European accent.[3]

The evidence for Shylock as a comic villain is partly in the literary and dramatic traditions, which Shakespeare followed, that lie behind the character, and partly in certain generic and other considerations.[4] Romantic comedy, as Shakespeare

[1] See also Christopher Spencer, *The Genesis of Shakespeare's 'The Merchant of Venice'* (Lewiston, NY, 1988), 88–92.

[2] On the famous speech that begins 'Hath not a Jew eyes?' (3.1.55) see 'The Play', below.

[3] Since medieval mystery and miracle plays portrayed Judas with red beard and hair and a large nose, later stage-Jews followed suit: see Landa, *The Jew in Drama*, 11; Calisch, *The Jew in English Literature*, 73; and Edgar Rosenberg, *From Shylock to Svengali: Jewish Stereotypes in English Fiction* (Stanford, Calif., 1960), 22. A ballad published in 1664 by an old actor, Thomas Jordan, indicates that Shakespeare's Shylock continued this tradition: see E. E. Stoll, 'Shylock', in *Shakespeare Studies* (New York, 1927), 255, 271, and Toby Lelyveld, *Shylock on the Stage* (Cleveland, 1960), 11. The large nose was also characteristic of Pantaloon's make-up in the *commedia dell'arte*, a secondary source for Shylock: see 'Sources', below, and John R. Moore, 'Pantaloon as Shylock', *Boston Public Library Quarterly*, 1 (1949), 33–42 (cited by Spencer, *Genesis*, 97). Had he intended to give Shylock an identifiable accent, Shakespeare could have done so, as he does, for example, Doctor Caius in *The Merry Wives of Windsor*. Nevertheless, many actors persist in using a comic—usually middle-European—accent when portraying Shylock, even though Spanish—the language of Sephardic Jews—was the *lingua franca* of European Jews in Shakespeare's time.

[4] See Marion D. Perret, 'Shakespeare's Jew: Preconception and Performance', *SStud* 20 (1988), 261–8.

developed the genre, is not without its darker elements, as Hero's denunciation and assumed death in *Much Ado About Nothing* clearly demonstrate and as, in a play closer to the *Merchant*, some aspects of *A Midsummer Night's Dream* also reveal.[1] Into this side of romantic comedy falls Shylock's design against Antonio's life. But the comic element also includes Antonio's hairbreadth escape. Legalistic quibbling over the validity of the bond, or Portia's arguments opposing, is not of paramount concern: Shylock's defeat is another in a long series going back beyond Barabas's descent into the cauldron—an example of 'the biter bit', a joke Elizabethans loved almost as much as jokes about cuckoldry. As for Shylock's conversion, we need only note that it was accepted as the alternative to something that, sinfully, Shylock thought would be worse. It could have been regarded by Elizabethan audiences (unlike those since then) as evidence of Antonio's Christian charity to Shylock—a mercy, combined with his request that Shylock be spared from destitution, entirely consonant with Portia's exhortations to Shylock earlier in the trial scene. In this way, the shallowness of Shylock's Judaism contrasts strikingly with the depth of Antonio's magnanimity and, before his, the Duke's spontaneous charity.[2]

But is Shylock worth saving? Apart from the consideration that every human soul is precious, does Shylock earn any serious sympathy that may lead us to rejoice in his salvation—such as it is? In spurning Shylock, Antonio and others, particularly Graziano, simultaneously spurn both his business and his religion; for in their minds—as in most Elizabethans'—usury and Jewishness were interlocked.[3] They thus

[1] See Jan Kott, *Shakespeare Our Contemporary*, trans. Boleslaw Taborski (Garden City, NY, 1964), 212–19, and Jay L. Halio, 'Nightingales That Roar: The Language of *A Midsummer Night's Dream*', in D. G. Allen and R. A. White (eds.), *Traditions and Innovations* (Newark, Del. 1990), 137–49.

[2] Cf. Bernard Grebanier, *The Truth about Shylock* (New York, 1962), 291; Cooper, 'Shylock's Humanity', 121; and Alan C. Dessen, 'The Elizabethan Stage Jew and Christian Example', *Modern Language Quarterly*, 35 (1974), 242–3.

[3] Stage-usurers were not necessarily Jews, but stage-Jews were invariably associated with usury. See Rosenberg, *From Shylock to Svengali*, 27. Kirschbaum, 'Shylock and the City of God', 25, and Warren D. Smith, 'Shakespeare's Shylock', *SQ* 15 (1964), 193–9, try (I think unsuccessfully) to distinguish between *ethnic* and *ethic* in Antonio's attitude toward Shylock.

provide Shylock with his deep resentment and the motivation for his revenge. Heaping injury upon insult, Jessica's elopement with Lorenzo, accompanied by bags of ducats and jewels, further exacerbates Shylock's bitter resentment. When news that Antonio's ships have miscarried reaches him at the moment of his agony over Jessica's actions, Shylock is more than ever prepared for a vicious counter-attack. His intention to hold Antonio to his bond and its penalty comes precisely when Salerio and Solanio taunt him unmercifully (3.1.21–49). True, Jessica later claims that Shylock always meant to undo Antonio if he could (3.2.282–8), but in the dramatic structure of the play, the impulse to revenge comes just here, where it is most powerfully motivated.[1] And it is through this motivation and the circumstances immediately surrounding it as much as anything else that Shylock's essential human quality emerges, just as Hamlet's action in the prayer scene—his lust for vengeance against Claudius—makes him not nobler but more human.[2]

Our response to Shylock, then, must be measured accordingly. Norman Rabkin is among the few critics who remind us that the scene with Salerio and Solanio involves the audience in a congeries of emotions so complex and contradictory that it is impossible to maintain a simple, single response to Shylock's behaviour. At once humorous, pathetic, antagonizing, Shylock's reaction to the news of his daughter's elopement, her theft, Antonio's misfortune, Jessica's squandering of his prized possessions parallels the similar situation in the Boar's Head scene in *1 Henry IV* where Falstaff makes his *apologia pro vita sua* (2.5.421–86).[3] If we are true to our experience of character and events, then no simplistic, reductivist description can appear accurate. Moreover, in subsequent scenes, our response to Shylock will be affected, or it ought to be, by an understanding not only of his position, but of

[1] In *Comic Transformations in Shakespeare* (London, 1980), 130–1, Ruth Nevo develops this point.

[2] See Jay L. Halio, 'Hamlet's Alternatives', *Texas Studies in Literature and Language*, 8 (1966), 169–88.

[3] Norman Rabkin, 'Meaning and *The Merchant of Venice*', in *Shakespeare and the Problem of Meaning* (Chicago, 1981), 7; cf. also pp. 22–3. Danson makes a similar point without citing the Falstaff passage, pp. 135–6.

his frame of mind, including the kinds of emotion his experience generates. That Shylock is hell-bent, literally, upon his revenge against Antonio should then hardly surprise us. Everything considered, his attitude and actions appear those of a man seriously deranged by what he, rightly or wrongly, regards as an enormous injustice against him personally and, through him, the people he represents.[1] Is it any wonder, then, that Shylock remains intransigent, impervious to Portia's appeals to mercy in the trial scene?

In this and other ways (developed below), Shakespeare reveals his attitude toward Shylock. It is ambivalent, far more than Marlowe's attitude toward Barabas. But in neither author can we confidently proclaim an anti-Semitic bias that is more than abstract and traditional. For Marlowe, the machiavel was more significant than Barabas's Judaism, which merged with it. By contrast, in developing Shylock's character in depth, and endowing it with vivid attitudes and emotions, Shakespeare succeeded in creating a dramatic figure who arouses far deeper feelings than Barabas can. Whereas the one remains, first and last, a comic stage villain, however brilliant and quick-witted, the other transcends the type, shatters the conventional image with his appeal to our common humanity, and leaves us unsettled in our prejudices, disturbed in our emotions, and by no means sure of our convictions.

Sources, Analogues, and Date

Although a play called *The Jew* (*c.*1578) was a precursor and possibly influenced the composition of *The Merchant of Venice*, very little is known about it other than Stephen Gosson's remarks in *The Schoole of Abuse* (1579). There it is described along with another play, *Ptolome*, as 'shown at the Bull, the one representing the greediness of worldly choosers, and bloody minds of usurers; the other very lively describing how seditious estates . . . are overthrown, neither with amorous gesture wounding the eye, nor with slovenly talk hurting the

[1] Cf. Shylock's complaint to Tubal, 'The curse never fell upon our nation till now; I never felt it till now' (3.1.80–2).

ears of the chaste hearers'.[1] Whether or not Shakespeare knew this old play, which may have been lost to him as well as to us, his immediate source was very likely a story in Giovanni Fiorentino's collection of tales called *Il pecorone* ('The Dunce'). Modelled on Boccaccio's *Decameron* and written at the end of the fourteenth century, the collection was printed in Milan in 1558, but not translated into English in Shakespeare's day. Shakespeare may have read the tale in the original along with another possible source, Masuccio Salernitano's *Il novellino* (1476), which contains a parallel to the Jessica–Lorenzo subplot. Or he may have otherwise picked up the stories through conversation or discussion. Yet another possible source, Richard Robinson's translation of the *Gesta romanorum* in 1595, contains an analogous casket story from which Shakespeare apparently borrowed the word *insculpt* at 2.7.57.[2]

Most of the major plot elements as well as themes in *The Merchant of Venice* appear in Fiorentino's first tale told on the fourth day in *Il pecorone*, though Shakespeare alters them and, combining them with elements from other sources (not least his own fertile imagination), transforms them into something far more compelling, both dramatically and intellectually. In Fiorentino's tale, for example, Giannetto (the hero) is the adopted son of a wealthy Venetian, Ansaldo, friend of his lately deceased father—not Giannetto's older friend, as Antonio is to Bassanio. Ansaldo showers gifts as well as affection upon Giannetto, and when the young man wants to take a trip with two of his friends to Alexandria, Ansaldo provides him with a ship like theirs, fully fitted out for the journey. En route, as the ships pass Belmonte, Giannetto learns about the mysterious lady who rules there. She has promised to give herself and everything she has to the man who possesses her.

---

[1] Cited from Arber's edition in Bullough, i. 445–6. By using a variety of approaches, including secondary sources and an analysis of the *Merchant*, S. A. Small has attempted to reconstruct the general outlines and essential substance of *The Jew* in ' "The Jew" ', *Modern Language Review*, 26 (1931), 281–7; but the results remain speculative. Compare M. A. Levy, 'Did Shakespeare Join the Casket and Bond Plots in *The Merchant of Venice*?', *SQ* 11 (1960), 388–91.

[2] It appears nowhere else in Shakespeare, although 'insculpture' occurs in *Timon*, 5.5.68. See Brown, pp. xxxii, 173.

Intrigued, Giannetto gives his friends the slip, enters the harbour, and is entertained graciously by the lady. In the evening, after a great feast, he and the lady retire to her bedchamber, where she invites him to drink before going to bed. The drink is drugged, and as soon as the young man's head touches the pillow, he falls asleep till well after daybreak. As a forfeit, Giannetto is obliged to surrender his ship and everything in it.

Giannetto sneaks back to Venice, chagrined and ashamed, but is at last persuaded to see Ansaldo, who believes his false story about shipwreck and forgives him. Smitten with love for the lady of Belmonte, Giannetto determines to try his luck again, and again he fails in exactly the same way. By now Ansaldo has spent most of his wealth, but he is willing to borrow from a Jewish money-lender so that Giannetto can try once more to sail to Alexandria and make his fortune, which he repeatedly proclaims as his goal. This time, taking pity on the handsome and debonair young man, a servant girl warns him about the drink, and Giannetto successfully possesses the lady—to everyone's delight and satisfaction. In his new-found joy, however, he forgets the due date of Ansaldo's bond, which elapses, leaving his benefactor at the mercy of the Jew, who thereupon demands his pound of flesh, the forfeit agreed upon. When belatedly Giannetto arrives with the money, the Jew refuses payment, insisting on his forfeit. Meanwhile, the lady also arrives, disguised as a lawyer, and foils the Jew's wishes. She also succeeds in getting a ring from Giannetto that his bride had given him, and at the end reveals her deception and cleverness. Ansaldo returns with them to Belmonte and is married to the servant girl who had told Giannetto how to win the lady.

While certain similarities to the Belmont plot in *The Merchant of Venice* are immediately apparent, differences also emerge that are equally if not more important. The lady's motivation for tricking her suitors suggests a concupiscence that does not fit with her otherwise noble character; hence, Shakespeare adds a more virtuous motivation, borrowed from the story in the *Gesta romanorum* (see below). The Jew's behaviour also lacks adequate motivation. Of an ancient grudge between the two principals, or a 'merry bond', there is nothing. The only suggestion of religious rivalry occurs in the Jew's explanation of his intransigence, simply that he could then boast that he had put to death the greatest of Christian merchants ('e'l Giudeo non volle mai, anzi voleva fare quello homicidio, per poter dire che avesse morto il maggiore mercantante che fosse tra' Cristiani').[1] While other merchants appeal to the Jew to relent, and Giannetto arrives and offers many times over the original sum borrowed, the trial lacks Portia's eloquent speech on mercy as well as most of the drama of the situation, though it heightens the comedy. Retaining the lady's insistence that the Jew extract precisely one pound of flesh without spilling so much as a drop of blood, the episode ends not with the forced conversion of the Jew, but with his furious anger as, outwitted completely, he tears the bond to pieces.

The Jessica–Lorenzo subplot and Graziano's courtship of Nerissa, with its doubling of the ring business, are also missing from this tale. So are the appearances of other suitors, counterparts to the Prince of Morocco and the Prince of Arragon. In adapting Fiorentino's story, Shakespeare did more than complicate his plot; he deepened its significance and broadened its themes. The introduction of the Jessica story, for example, suggested by Marlowe's *Jew of Malta* and Masuccio's tale, not only strengthens Shylock's motive for revenge, as Muir says;[2] it also adds to the comic conclusion of three happily married couples. But here again Shakespeare alters his sources; for Antonio, unlike Ansaldo, is left at the end as at the beginning a 'fifth wheel'. Although his ships, once believed wrecked, have miraculously returned (5.1.276–7), he may have as much cause for melancholy as at the start: he has just narrowly escaped death and, his dearest friend now married, he faces a lonely future. Touches of mortality, apparent though seldom emphasized in early Shakespearian comedy (most notably in *Love's Labour's Lost*), round off the play in ways found neither in Shakespeare's sources nor in the related work of his contemporaries.

---

[1] *Il pecorone di Ser Giovanni Fiorentino* (Venice, 1565), 40; reprinted in slightly different form in J. Payne Collier, *Shakespeare's Library* (London, n.d.), II. ii. 86.

[2] Kenneth Muir, *The Sources of Shakespeare's Plays* (New Haven, Conn., 1977), 90.

The Casket Story

The flesh-bond story, which Shakespeare fused with the casket story, has its origin in tales as ancient as one in the *Maha-bharata*, where King Usinara saves a dove from a hawk by giving it some of his own flesh instead. The Talmud includes a similar story of sacrifice involving Moses, and under the Twelve Tables of Roman Law, creditors were allowed as a last resort to claim the body of a defaulting debtor and divide it among them.[1] Although the first story in English may be found in the *Cursor mundi*, dating from the thirteenth century,[2] Shakespeare's use of the casket story probably derives from History 32 of the *Gesta romanorum*, as translated by Richard Robinson, but it too has ancient origins in myth and fable as well as great psychological and symbolic significance. In Robinson's version of the *Gesta romanorum*, Ancelmus, the emperor of Rome, agrees to marry his only son to the daughter of the king of Ampluy. After undergoing many hardships, including shipwreck at sea, the princess arrives at the emperor's court, but she must pass a further test before she can marry the emperor's son. The emperor sets before her three caskets of gold, silver, and lead. The first, encrusted with precious stones but containing dead men's bones, is inscribed, 'Whoso chooseth me shall find that he deserveth'. The second, crafted of fine silver but containing earth and worms, is inscribed 'Whoso chooseth me shall find that his nature desireth'. The leaden casket, inscribed with the motto 'Whoso chooseth me, shall find that God hath disposed for him', is filled with precious stones. The princess piously chooses the last one, to the satisfaction of all concerned, and is married to the prince.[3]

Borrowing the casket story, Shakespeare obviously did more than alter the inscriptions. He dramatized the selection of each casket and, in fusing the tale with the flesh-bond plot, made the choosers men. In his study of 'the Theme of the Three Caskets', Sigmund Freud speculated on the significance of the

[1] Bullough, 446–7.

[2] The story, which has a Christian goldsmith as the borrower and a Jewish money-lender as his creditor, is paraphrased and summarized by L. Toulmin Smith in Furness, 313–14.

[3] See the relevant passages in Brown, Appendix V, pp. 172–3.

story as Shakespeare adapted it not only in *The Merchant of Venice* but also in *King Lear*.[1] While he is more deeply interested in the tragedy, he makes some important observations about the comedy as well. He begins by noting the relation of the story to an astral myth, identified by Eduard Stucken, who saw the Prince of Morocco as the sun, the Prince of Arragon the moon, and Bassanio the star youth (*Astralmythen der Hebraer, Babylonier und Aegypter*, Leipzig, 1907). But Freud goes further. The astral myth is, after all, a projection of a human situation, and it is the human content of the myth that is important.

The change from female to male choosers in *The Merchant of Venice* is clearly a significant inversion of the theme. Symbolically, the caskets—like coffers, boxes, or baskets—are equivalent to women. Hence, the choice involved is of three women, as in the Greek legend of the Judgement of Paris. The third woman, the one Paris chooses, is Aphrodite, the beautiful Goddess of Love. But the symbolism of the leaden casket hardly points in this direction; it points towards death. If the third goddess is the Goddess of Death, then all three goddesses are the Fates, the Moirai, the third of whom is Atropos. Yet no one consciously chooses death; one falls victim to it, inexorably.

Freud resolves this apparent contradiction through his theory of 'reaction-formation', which argues that human beings substitute for the terrors of reality a more acceptable version, one that satisfies their wishes as reality often does not ('wish-fulfilment'). Human nature thus rebelled against the myth of the Moirai and imaginatively constructed a counter-myth, one derived from it, 'in which the Goddess of Death was replaced by the Goddess of Love and by what was equivalent to her in human shape. The third of the sisters was no longer Death; she was the fairest, best, most desirable and most lovable of women' (p. 299). Through similar wishful reversal, choice entered the myth of the three sisters, or Fates, and became a substitute for necessity, or destiny:

---

[1] *The Standard Edition of the Complete Psychological Works of Sigmund Freud*, trans. James Strachey (London, 1958), xii. 291–301.

In this way man overcomes death, which he has recognized intellectually. No greater triumph of wish-fulfillment is conceivable. A choice is made where in reality there is obedience to a compulsion; and what is chosen is not a figure of terror, but the fairest and most desirable of women. (Ibid.)

Bassanio's 'choice', then, becomes his fate, and he and Portia, the Aphrodite figure, are destined for life together in love. But first Shylock, the killjoy, the true spirit of deathliness, must be overcome. Like the princess in the original casket story, or rather the clever lady in the flesh-bond tale, Portia accomplishes this deed. Thus the two stories are fused.[1]

The Jessica–Lorenzo Subplot

The fourteenth story in Masuccio's *Il novellino* provides the likeliest source for the Jessica–Lorenzo subplot, although Abigail, Barabas's daughter in Marlowe's *The Jew of Malta*, may have first led Shakespeare to think of a daughter for Shylock. In Masuccio's tale, an old miser, a merchant of Messina, has a lovely young daughter much like Miranda in *The Tempest*, someone who has hardly ever seen a man. She falls in love with Giuffredi Saccano, a cavalier, who notices her one day at her window and is likewise inflamed with passion for her. The cavalier plots with a slave girl to trick the old man and get Carmosina to elope with him. He succeeds; moreover, Carmosina steals away with a great store of her father's treasure, leaving the miser doubly bereft.[2]

A related situation occurs in Book III of Anthony Munday's prose tale, *Zelauto or The Fountaine of Fame* (1580), which bears other similarities to Shakespeare's play, including a flesh-bond story. There, two friends, Strabino and Rodolfo, woo two young women, Cornelia and Brisana. Cornelia is also sought by an old miser named Truculento, who is Brisana's father. To ensure success in their suits, the two young men

---

[1] Norman Holland goes further and relates Freud's idea to the play as a whole through 'the theme of venturing, which links the romantic plot to the mercantile one. This third woman, death (lovely, rich, and merciful in a Christian view), stands for the investor's return in the great venture of life itself', 'Freud on Shakespeare', PMLA, 75 (1960), 171.

[2] The story is reprinted from W. G. Waters's translation (1895) in Bullough, 497–505, and in Joseph Satin, *Shakespeare and his Sources* (Boston, 1966), 142–9.

borrow a sum of money from Truculento, pledging to repay it on a certain date or else forfeit their right eyes. Strabino and Rodolfo win their suits, much to the furious displeasure of Truculento, who then seizes on the missed payment of their debt to claim the forfeit. The immediate spur to his revenge is thus similar to Shylock's. Furthermore, in a trial before the magistrate, Cornelia and Brisana pose as lawyers and successfully plead their fiancés' case. They win partly because Truculento (again like Shylock) has failed to claim as payment any drop of blood along with the right eyes of the debtors. Truculento is denied his principal, but eventually accepts Rodolfo as his son-in-law, making him his heir.[1] Missing from both Masuccio's tale and Munday's, however, is the motif of a Jewess's love for a Christian and her subsequent conversion, which Shakespeare must have seen in Abigail's relationship with her lover in Marlowe's play and adapted accordingly.

Other possible sources and analogues include the ballad of *Gernutus*,[2] Gower's *Confessio Amantis*, Book V,[3] and Alexandre Sylvain's *The Orator*, which was translated from the French by 'L.P.' (Lazarus Piot, a pseudonym Munday used) and published in 1596.[4] Of these, the last is the most important, for Declamation 95, 'Of a Jew, who would for his debt have a pound of flesh of a Christian', contains the shape of arguments very similar to those Shylock and others use in the Trial Scene. For example, Sylvain's Jew, declining to say specifically why he prefers the pound of flesh to the money owed, offers various possible reasons, although they are quite different from the more caustic response Shylock gives at 4.1.39–61. Indeed, comparison between Shakespeare's play and its sources shows, here as elsewhere, a far more fertile

---

[1] Compare Bullough, 452–4, and Brown, p. xxxi, who reprint excerpts from *Zelauto* on pp. 486–90 and 156–68, respectively.

[2] Reprinted in Brown, 153–6. The date is uncertain but is roughly contemporary with the *Merchant*, which may actually have been its source instead of the other way round. Bullough, 449–50, believes the ballad is pre-Shakespearian and may derive from *The Jew*.

[3] Excerpted in Bullough, 506–11.

[4] Reprinted in Bullough, 482–6; Brown, 168–72; Satin, *Shakespeare and his Sources*, 138–41; Furness, 310–13. Compare Muir, *Sources*, 87. In '"Lazarus Pyott" and Other Inventions of Anthony Munday', *Philological Quarterly*, 42 (1963), 532–41, Celeste Turner Wright ascertains that 'L.P.' was, in fact, Anthony Munday.

imagination at work, able not only to refashion and transform adapted materials, but by dramatizing them afresh in striking and vivid language, to recreate them almost as if they were new.

Biblical and Classical Allusions

The Merchant of Venice is studded with both classical and biblical allusions, though how many were deliberately intended or unconsciously woven into the fabric of the dialogue is uncertain. Similarly, the degree to which Shakespeare could depend upon audience recognition is debatable. Some, of course, cannot be mistaken or missed, as in Shylock's long account of Jacob, Laban, and the parti-coloured lambs (1.3. 68–93), or the repeated references to the story of Jason and the golden fleece (1.1.169–72, 3.2.240, etc.). Others may be more subtle, involving satirical or other purposes.[1] But all of them extend the play's dimensions, helping to universalize its themes and providing a broader context of human experience.

More controversial is the extent to which Shakespeare may have been consciously building allegorical elements into his play, adapting to his secular theatre the more forthright examples of morality and miracle plays that he would have witnessed as a youth in Warwickshire and that still had an important influence on the drama of the 1590s, as in Marlowe's *Dr Faustus*. Lancelot Gobbo's debate on whether to leave his master the Jew (2.2.1–29) is a comic interlude adapted from moralities that present good and evil angels arguing with an Everyman character on how to behave. This much is clear. But how deliberately Shakespeare intended an allegory of the Old Law versus the New in the trial scene of Act 4 is less certain. Though allegorical elements may be identified—Shylock 'stands for' law, Portia advocates mercy— the characters are not simply embodied abstractions typical of that mode (see below). The episode may owe something to the *Processus Belial*, a medieval story in which the Devil pleads for mankind's soul in the court of heaven, basing his appeal on strict justice. The Virgin Mary, however, intercedes, claiming that mercy is also an attribute of God, and succeeds in

---

[1] See e.g. commentary on 2.1.32–8; 4.1.196–8, 203; etc.

her appeal. During the debate, the Devil even produces a set of scales to weigh the part of humankind that is owed him. Shylock's behaviour and his identification with the Devil in *The Merchant of Venice* brings the analogue to mind, especially as the plea for mercy and other details do not appear in any of the recognized sources, but how aware of it Shakespeare was during composition is impossible to determine.[1]

On the other hand, Shakespeare apparently read the Bible (in the Geneva translation) to learn something about Hebraic traditions, specifically to help him develop the character of Shylock, since he evidently knew little at first hand about Jews or Jewish traditions.[2] Besides Shylock's long account of Jacob and Laban, there are allusions to Rachel's trick in obtaining Esau's birthright for her other son, to Hagar and her offspring, to a synagogue where Shylock takes his oath, to Jewish dietary restrictions, and so forth. Shylock's name may also derive from biblical sources, although its derivation is less certain than the names of other Jewish characters.

In the genealogies of Genesis 10–11 Shakespeare found the Hebrew names he needed. Tubal is there (Gen. 10:2) and Cush, Ham's first-born son and the father of Nimrod (Gen. 10:6–9; spelled 'Chus' in the Bishops' Bible and others except Geneva). Iscah, the daughter of Abraham's brother Haran, appears a little later (Gen. 11:29); from it, the name Jessica derives.[3] All are descendants of Shem, Ham, or Japheth, the sons of Noah. So is Shelah, Shem's grandson, mentioned twice in these chapters (Gen. 10:24 and 11:12–

[1] See J. D. Rea, 'Shylock and the Processus Belial', *Philological Quarterly*, 8 (1929), 311–13. The Weighing of Souls appears in medieval wall paintings as well, and in morality plays, e.g. *The Castle of Perseverance*, similar debates occur (NS 9).

[2] See above, 'Shakespeare and Semitism'. The point is made by Mahood in an appendix on Shakespeare's use of the Bible in *The Merchant of Venice* (NCS 184–8). She also notes that Shakespeare, like many of his contemporaries, would be familiar with both the Bishops' Bible (1584 edn.) and the Geneva (1596). Although echoes from the former outnumber those from the latter, allusions to the Geneva's marginal glosses indicate that Shakespeare was probably reading that version as he wrote this play.

[3] J. L. Cardozo, *The Contemporary Jew in Elizabethan Drama* (Amsterdam, 1925), 226, notes that according to the great Hebraist Rashi, Iiskah was an early name of Sarai (or Sarah), wife of the patriarch Abraham and mother of Isaac. The Geneva marginal gloss makes the point as well. Compare Sinsheimer, *Shylock*, 87, who also discusses the derivation of the names.

15). Shelah begot Eber and thus became the progenitor of Hebrews.[1] Shylock's name probably derives from Shelah, in Hebrew 'Shelach', still closer in pronunciation to that of Shakespeare's character, whose first syllable probably contained a short *i*, lost in today's pronunciation.[2] His wife's name, Leah, occurs later in Genesis 29 and again in chapter 30, where the story of Jacob's dealings with Laban appears.[3]

Other Influences

No literature emerges from a vacuum, though the lines of descent for ideas, images, terms may be more or less direct or indirect. From all accounts, Shakespeare was a good reader; in addition, he must have been a good listener, picking up references to books, people, and events in conversation with his friends and colleagues almost daily. Because of the close proximity of the tales of Thisbe, Dido, and Medea in *The Legend of Good Women*, which follows closely upon the *Troylus and Creseyda* in old editions of Chaucer, some scholars believe Shakespeare may have had his Chaucer open before him as he began Act 5 (see commentary on 5.1.1–14). Less direct may have been the influence of Boethius' *De institutione musica* in the same act, when Lorenzo lectures Jessica on the music of the spheres.[4]

Shakespeare was well acquainted with the law—like many of his contemporaries, he was himself litigious—but discussions of equity and justice, based upon his assumed intimate

[1] As the marginal gloss to Gen. 10 : 24 in the Geneva Bible states.

[2] Cardozo, *Contemporary Jew*, 223, notes that *y* and *i* were interchangeable and suspects a pun, Scylla–Shylock, in Lancelot's speech, 3.5.14. Note also the Q1 spelling '*Shyloch*', when Bassanio addresses him, 1.3.49. Not everyone agrees with this etymology, however. For example, Maurice Brodsky conjectured a derivation from the phrase *shelee shelee v'sheloch sheloch*, representing a man who stood on the letter of the law. Israel Gollancz noted the association with *shallach*, the Hebrew word for cormorant, a synonym in Elizabethan English for usurer. And Norman Nathan, who cites the previous two contenders, argues instead for a cognate with *shullock*, an obsolete term of contempt which the *OED* records as early as 1603 ('Three Notes', 152–4). But given the biblical origin of the other Jewish names and their close association with Shem's grandson, the derivation from Shelah, or Shelach, seems more likely. Cf. Brown, 3; Merchant, 171; Spencer, *Genesis*, 96–7.

[3] For the ironic appropriateness of her name, see commentary on 3.1.114.

[4] See Danson, 187, who cites John Hollander, *The Untuning of the Sky* (Princeton, NJ, 1961), 25.

knowledge of Coke or William Lambarde's *Archaionomia* (1568), may be quite beside the point, dramatically speaking, as we try to understand what happens in the trial scene in Act 4.[1] Similarly, we recognize Shakespeare's knowledge of Roman comedy; *The Comedy of Errors*, one of his earliest plays, is adapted from a comedy by Plautus. But Shylock and Jessica may owe as much to the pantaloon and his daughter in the Italian *commedia dell'arte*, and in any case they derive more directly from Masuccio's tale. Much was written about usury and its evils during Elizabeth's reign, and doubtless Shakespeare was aware of current controversy. But did he actually read Calvin's arguments and Beza's in support of it, or Miles Mosse's *The Arraignment and Conviction of Usury* (1595) against it,[2] to understand the terms of the controversy? And how relevant are the details of usury to *The Merchant of Venice* after all? Although Shylock and Antonio debate lending money at interest in 1.3, Shylock expressly forgoes any return on his loan in his apparent attempt to be friendly.[3] From that point on, not interest on the loan but the security, or its forfeit, becomes the central issue.[4]

The 'Myth' of Venice

The so-called 'myth' of Venice, on the other hand, certainly exerted an influence in the composition of the play.[5] To Shakespeare and his contemporaries, Italy was in every sense romantic, and doubtless for that reason many of the dramas of the time, comedies and tragedies alike, were set there. Venice had a special appeal. An ancient republic, it had earned a reputation for political astuteness, great wealth, and legal justice. In addition, it was a pleasure-loving city, to

[1] See E. F. J. Tucker, 'The Letter of the Law in *The Merchant of Venice*', *SSur* 29 (1976), 93–101.

[2] Compare Brown, p. xliii. In ' "When Jacob Graz'd his Uncle Laban's Sheep" : A New Source for *The Merchant of Venice*', *SQ* 36 (1985), 64–5, Joan Ozark Holmer claims Mosse's book as a source, specifically for Shakespeare's use of the Jacob–Laban story.

[3] See 1.3.134–8 and the critical interpretation of the play below.

[4] Of course, as Brown notes (p. xliii), Antonio has been a chief obstacle to higher rates of interest among Venetian usurers, and this is an important motive underlying Shylock's decision to insist upon the forfeiture (3.1.119–21). But revenge for all the wrongs he feels he has suffered is more important.

[5] In this paragraph and the next, I am indebted to McPherson, esp. ch. 2.

1. View of Venice, from *Civitatis Orbis Terrarum* (1593), by Georg Braun and Frans Hogenburg

which swarms of tourists flocked constantly, including many from England.[1] Shakespeare could have become aware of these aspects of the myth in a variety of ways—through both oral and written sources—though we have no evidence that he himself ever visited there.

Of especial importance to *The Merchant of Venice* was the city's reputation for far-flung trade; if Shakespeare exaggerates its extensiveness,[2] which by then was in decline, her reputation as a great maritime power nevertheless was commonly accepted. So was her reputation for fairness in dealing with foreigners, which had contributed to making her a great maritime power. Her vaunted impartiality was seconded by the severity of her justice. Fynes Moryson, who journeyed to

[1] Among the most famous of these is Thomas Coryat, who visited Venice in 1608. His account of the visit (including his encounters with Jews) is in *Coryat's Crudities Hastily gobled vp in fiue Moneths trauells in France, Sauoy, Italy . . .* (1611).

[2] See commentary on 3.2.266.

2. Venetian carnival, from Giacomo Franco, *Habiti d'huomini et donne Venetiane* (1609)

Venice in 1593–4, saw two young sons of senators mutilated and then executed for singing blasphemous songs and for other misdemeanours.[1] Speaking of the Venetians, Lewis Lewkenor notes 'the greatness of their Empire, the gravity of their prince, the majesty of their Senate, the inviolableness of their laws, their zeal in religion, and lastly their moderation and equity'[2]—qualities exhibited by characters and events in Shakespeare's play. In recalling, too, 'the delicacy of their entertainments, the beauty, pomp, and daintiness of their women, and finally the infinite superfluities of all pleasure and delights', he pays homage to the city as the pleasure capital of Europe, an aspect of Venice also revealed in *The Merchant of Venice* and later cited in *Othello*.

Though discriminated against in various ways, Jews were tolerated in Venice and allowed to practise their religion openly,

[1] McPherson, 37.
[2] *The Commonwealth and Government of Venice* (1599); quoted by McPherson, 39.

26

as they were not in London or anywhere in England. Compelled to wear distinctive garb,[1] to live in a special district called the 'ghetto' (the Italian name for a foundry that was once built there), and to pay disproportionately heavy taxes, they could still earn a living; in fact, partly because so few other occupations were open to them, they became the city's leading moneylenders and second-hand dealers, or pawnbrokers. Relations between Christians and Jews were far from cordial, but animosity was based upon religious grounds, not racial. Shakespeare knew or understood some of these details, though by no means all. Shylock, for example, apparently lives in Venice proper, not in the ghetto, which is never referred to by name. But Antonio explicitly invokes the reputation for impartial justice and its importance for foreign trade, the thought of which makes him despair of help from the Duke (3.3.26–31).

Date

Although not printed and published until 1600, *The Merchant of Venice* was entered in the Stationers' Register on 22 July 1598 (see below, p. 85). Six weeks later, on 7 September, Francis Meres's *Palladis Tamia* was also entered on the Register. In this book, Meres specifically mentions *The Merchant of Venice* among Shakespeare's comedies. The references clearly establish the last possible date for the play.

How early it was written is more difficult to determine. Salarino's reference to 'my wealthy Andrew docked in sand' (1.1.27) apparently alludes to the San Andrés, a Spanish galleon that, along with the San Matias, ran aground and was captured fully loaded by English ships under the earl of Essex in Cadiz harbour, June 1596.[2] The ship was later renamed the Saint Andrew and had further misadventures the year following, so that it was much in the news during the

---

[1] Shylock's 'Jewish gaberdine' (1.3.109; see commentary) may be Shakespeare's way of indicating this fact, although actually a still more distinctive and humiliating garb was worn, as in the RSC production of 1984. This consisted of a badge (a yellow circle) and a yellow or red high pointed hat, or turban.

[2] Sir William Slingsby, 'A Relation of the Voyage to Cadiz', *The Naval Miscellany I*, ed. J. K. Laughton (1902), 25–92 (NS I). Compare Brown, p. xxvi, who notes that Ernest Kuhl first identified the allusion in *TLS*, 27 Dec. 1928, p. 1025.

period 1596–7. On her return from the Islands Voyage, for example, after much battering by storms, Essex decided not to risk her sailing by the Goodwin Sands, mentioned in *The Merchant of Venice* as a graveyard for ships (3.1.4–5).

Two other pieces of external evidence also suggest 1596–7 as the approximate date for composition. Marlowe's *The Jew of Malta* was still being performed by the rivals of Shakespeare's company, the Lord Admiral's Men; in 1596 it was performed eight times (Brown, p. xxiv), and *The Merchant* was doubtless designed to capitalize on its popularity. Since the Lord Chamberlain's Men had lost their lease at the Theatre and the Globe had not yet been built, they were playing at this time at the Swan on the Bankside, near where the Globe would eventually stand. To obtain the right to perform in that theatre, members of Shakespeare's company were compelled by its proprietor, Francis Langley, to sign 'outrageously exorbitant' bonds. These stipulated that the players would perform exclusively at the Swan, on forfeit of the huge sum of one hundred pounds. Bonds were a new form of contractual agreement in the theatre; Philip Henslowe soon afterwards resorted to their use as well, and Shakespeare's company became involved with his requirement for a bond when the Swan was later refused a licence to open. Although the terms of the bond are unknown, the negotiations very likely influenced Shakespeare's thinking as he wrote the *Merchant*.[1]

Internal evidence, based upon stylistic analysis, also suggests 1596–7 as the date of composition. According to the editors of the Oxford *Complete Works*, Oras's pause test links it to the *Henry IV* plays, but the handling of the imagery and verse is less mature; hence, in their chronology it appears after the 'lyric' group (*A Midsummer Night's Dream*, *Romeo and Juliet*, *Richard II*), and between *King John* and *1 Henry IV*.[2] This ordering makes a good deal of sense. The verse, especially in Act 5, strikingly resembles that of the lyric plays, but the

---

[1] See James Shapiro, 'Which is *The Merchant* here, and which *The Jew*', *SStud* 20 (1988), 275–6, who cites (among others) William Ingram, *A London Life in the Brazen Age: Francis Langley, 1548–1602* (Cambridge, Mass., 1978) and C. W. Wallace, 'The Swan Theatre and the Earl of Pembroke's Players', *Englische Studien*, 43 (1911), 340–95.

[2] *TC* 119–20.

prose passages, such as Shylock's 'Hath not a Jew eyes' (3.1.55–69), look forward to Falstaff's.[1] As the first of Shakespeare's great comic heroines, Portia was soon followed by Beatrice, Rosalind, and Viola, with whom she shares a number of traits. Similarly, Bassanio, the handsome but somewhat feckless romantic hero, anticipates Claudio in *Much Ado* and Orlando in *As You Like It*. But in the treatment of comic villains, Shakespeare must have learned from the development of Shylock's character a need for some restraint. He obviously exercised a great deal as regards Don John, just the right amount in the creation of Malvolio.

The Play

The emphasis upon different kinds of bondage and bonding in *The Merchant of Venice* opens the play to a comprehensive critical interpretation that takes account of inconsistencies and contradictions without reducing actions, characters, and events to an oversimplified, unified pattern.[2] In his very first lines Antonio confesses the grip that melancholy has upon him, and shortly thereafter we see the bonds, both weak and strong, that tie him to friends and acquaintances. The bond of friendship with Bassanio involves debts of loyalty but also money that almost immediately lead to other debts, obligations alternatively despised, feared, or lightly entered into, when Bassanio approaches Shylock in Antonio's name to secure a sizeable loan. But before Shakespeare presents that confrontation, Portia and Nerissa at Belmont debate other obligations, those involving the bond between parent and child, father and daughter. By the time the second act concludes, still other bonds between parents and children are affirmed or broken, as are loyalties to religion, master-servant relationships, holiday sport—even time and tides.[3] Through it all, and indeed throughout the entire action, the play both implicitly and explicitly affirms the bondage that our common humanity imposes, not only between individuals but also as

[1] Cf. e.g. *1 Henry IV* 2.4.471–85, 5.1.127–40.

[2] See above, 'Shakespeare and Semitism.'

[3] When the wind has 'come about', Bassanio and his entourage have to leave before the masque can begin in Act 2: see 2.6.62–8.

3. Italian Merchant, from Cesare Vecelli, *De gli habiti antichi et moderni . . .* (1590)

MERCANTI.

GIOVA. NETTI.

4. Italian Young Man, from Cesare Vecelli, *De gli habiti antichi et moderni . . .* (1590)

regards our 'muddy vesture of decay'. Important, too, is the hold that great dramatic art has upon its audience—and the reasons for that fascination.

Although Granville-Barker maintains that *The Merchant of Venice* is 'a fairy-tale' (p. 67), it does not begin like one, and other critics have noticed the play's mixture of realism and convention, or artifice and naturalism.[1] Antonio's melancholy has been variously explained, but *in the play* all the explanations that his friends offer are rejected. Like many people who suffer from free-floating depression, Antonio does *not know* why he is 'so sad'. It troubles him as much as, or more than, others; and whatever the cause, its grip appears relentless. By the end of the play, although his miraculously returned ships (part of the 'fairy-tale' quality of the action) have given him 'life and living' (5.1.286), he is scarcely cured of his malady. Apparently a seventh wheel among the newly-wedded couples,[2] he is cordially welcomed into Portia's house; but what he will do there or what his role will be now that Bassanio is safely married and no longer needs his friend as he did in 1.1 remains in question. Since the play ends at this point, no answer is forthcoming. That is one way (though not the most important way) the play retains its fascination for us.[3]

Psychoanalytically oriented critics and stage directors have explained Antonio's melancholy as the result of his homosexual attachment to Bassanio, his young friend. The theory is plausible but unprovable.[4] Without doubt, Antonio's attachment to Bassanio is as strong as it is real, as the account of their leave-taking (2.8.36–49) demonstrates; but that it goes beyond a Platonic attachment between friends cannot be determined. While homosexual love in the Platonic hierarchy

[1] For example, Danson, 92–3; Leggatt, 119–22.

[2] For Shakespeare's alteration of his source, see 'Sources', above.

[3] For five possible ways of ending the play, see Thomas Clayton, 'Theatrical Shakespeare regresses at the Guthrie and Elsewhere: Notes on "Legitimate Production" ', *New Literary History*, 17 (1985–6), 331–3, and cf. Leggatt, 149.

[4] See Holland, 238–9, 331, and compare Graham Midgley, '*The Merchant of Venice*: A Reconsideration', *Essays in Criticism* 10 (1960), 125: 'Antonio is an outsider because he is an unconscious homosexual in a predominantly, and indeed blatantly, heterosexual society'. See also Danson, 34–40.

rests on a higher plane than heterosexual love, it is by no means the highest level.[1] Nor can we confidently attribute Antonio's melancholy to a version of *Weltschmerz*, although quite possibly his affliction derives from a world-weariness sometimes experienced in middle life. Whatever the cause (very likely more than one), he appears ready to sacrifice himself for his friend. As 'a tainted wether of the flock' (4.1.113), he sees himself as a scapegoat figure, though what sins he carries on his back remain, like the causes of his melancholy, unspecified.

On his part, Bassanio is also tied to Antonio, by more than bonds of friendship. He has, he acknowledges, borrowed from Antonio before, but like a true gentleman Antonio refuses to let his friend dwell on that fact (1.1.153–4). Keen as his needs are, and anxious as he is to try to win the lady of Belmont, Bassanio tries to dissuade Antonio when it actually comes to signing the agreement with Shylock. Although some critics see him as little other than a self-aggrandizing opportunist,[2] others view Bassanio as a naïve young man who, like many of the characters, has much to learn about the ways of the world as well as himself. As the play progresses, he proves an apt pupil, bound to learn.[3]

Bound, also, is Portia—to her father's will as against her own desires in the way of matrimony. If the weariness she voices in her first appearance echoes Antonio's, we hardly need Nerissa's explanation to learn the cause of it. Like any rich, spoiled heiress, she is bored with her lot; surfeiting with too much, she feels as bad as those who suffer from too little. Until Bassanio comes along and releases her from her enclosed world—at least temporarily—into 'the world', she has hardly anything to do. Like Rosalind and Celia at the beginning of *As You Like It*, or the Brangwen sisters in the opening chapter of *Women in Love*, she therefore talks with Nerissa about love and marriage, bridling at her father's decree concerning the choice of a husband, but determined nevertheless to observe it. She thus stands in direct contrast to Jessica later, another

---

[1] See Plato, *The Symposium*.

[2] Sir Arthur Quiller-Couch, for example, calls Bassanio a 'predatory young gentleman' (NS, p. xxv).

[3] See Danson, 110–11.

only child, who breaks the bonds to her father and to her religion in eloping with Lorenzo, compounding her double disloyalty by stealing ducats and jewels from Shylock as well.

Early in Act 2 Lancelot Gobbo also focuses on family ties, though he begins with a broadly comic soliloquy on breaking the bond to his master Shylock in search of better employment. His dilemma raises questions about his 'honesty', leading eventually to a coarse joke concerning his father's marital fidelity, a joke he amplifies when Old Gobbo appears ('it is a wise father that knows his own child', 2.2.72–3). Insisting on calling himself 'Master', Lancelot not only confuses and embarrasses his father, but satirizes upward social mobility, an oblique glance, perhaps, at Bassanio's wish to become lord of Belmont by winning Portia—and possibly a self-reflexive joke on Shakespeare's own social ascendance at this time.[1]

Lancelot does more than comically abuse his filial relationship and justify breaking with Shylock, his master, to find new service with Bassanio. His frequent malapropisms and other examples of verbal pretentiousness comically bend, if they do not break, the boundaries of ordinary speech.[2] He demonstrates his legitimate descent from Old Gobbo, linguistically and otherwise, in their address to Bassanio:

GOBBO He hath a great infection, sir, as one would say, to serve—
LANCELOT Indeed, the short and the long is, I serve the Jew, and have a desire, as my father shall specify—
GOBBO His master and he, saving your worship's reverence, are scarce cater-cousins—
LANCELOT To be brief, the very truth is that the Jew, having done me wrong, doth cause me, as my father—being, I hope, an old man—shall frutify unto you— (2.2.120–8)

Later, Lancelot's linguistic virtuosity, such as it is, appears in dialogue with Jessica, when he confesses his 'agitation' regarding her putative damnation for being a Jew's daughter (3.5.1–23). Trying to comfort her, he ends by compounding her predicament, as he sees it, twice over: first, by holding

[1] See S. Schoenbaum, *William Shakespeare: A Compact Documentary Life* (Oxford, 1987), 227–30, concerning the grant of a coat of arms to Shakespeare's family in 1596, about the time this play was written.

[2] On linguistic perversions in the play, Lancelot's and others', compare Danson, 96–104.

out the hope that Shylock did not father her after all—a kind of 'bastard' hope, as both agree; then by rejecting her salvation at the hands of her husband, Lorenzo, who by making her a Christian will raise the price of pork, causing widespread shortages. When Lorenzo enters, Lancelot changes tactics and engages in other forms of chop logic and verbal quibbling, to Lorenzo's near exasperation: 'How every fool can play upon the word' (40). The word-play, comic here, ironically foreshadows the linguistic display in Act 4 that ends with another conversion—Shylock's.[1]

Different kinds of verbal pretentiousness also characterize Portia's suitors, particularly the Prince of Morocco. He opens Act 2 with a self-conscious and extravagant defence of his complexion, 'The shadowed livery of the burnished sun' (2.1.2). Bragging of his red blood and of his military and sexual accomplishments, he confuses classical allusions, collapsing the story of Hercules playing at dice into the quite different story of the poisoned shirt of Nessus. Similarly, in his boasts of military prowess, he falsifies history (see commentary on 2.1.25–6 and 32–8). Withal, he is too dense to pick up Portia's hint (doubtless unintentional, or if not, then a test of her suitor's perspicacity) when she disdains the appeal of visual attractions (2.1.13–14). When the time comes for the actual choice of caskets, he rejects the leaden one, indicating his predisposition with the words, 'A golden mind stoops not to shows of dross' (2.7.20). The silver casket appears attractive, especially as it encourages a choice based upon self-worth, of which Morocco has plenty. But the golden one seduces him with 'Who chooseth me shall gain what many men desire' (37), whereupon the prince breaks into hyperbole praising 'the lady'—and indirectly himself for 'solving' the riddle. His reward is the death's head and a scroll that reads, in part:

> Had you been as wise as bold,
> Young in limbs, in judgement old,
> Your answer had not been inscrolled.
>
> (70–2)

---

[1] Leggatt, 140.

The words anticipate not only Bassanio's correct choice later, but Portia's actions in the trial scene, where she resembles a 'Daniel come to judgement' (see 4.1.220, 247, and commentary).

Arragon's pretentiousness is of another sort. Rejecting the golden casket, because he 'will not jump with common spirits' (2.9.31) who are thus attracted ('what many men desire'), he 'assumes' desert (50) and chooses the silver one after a long and pompous deliberation on the matter. His pedantry is rewarded with a fool's portrait and words that say:

> There be fools alive, iwis,
> Silvered o'er, and so was this.
> Take what wife you will to bed,
> I will ever be your head.
>
> (67–70)

After his swift departure, Portia comments on 'these deliberate fools' who in choosing 'have the wisdom by their wit to lose' (80), or in other words, who outsmart themselves.

The inconsistency in the scroll's words—suggesting that Arragon is not bound to his pledge never 'To woo a maid in way of marriage' (13)—has been variously explained (see commentary) but is only one among the play's many inconsistencies and contradictions. It differs from the ambiguities of language that simultaneously bind and unloose, as Macbeth finds and in this play Shylock will—to his cost. Another link between the casket plot and the flesh-bond plot is the oath that the suitors take before making their election and the confidence they exhibit in themselves. Their hubris and Antonio's, as he agrees to the 'merry bond' and its diabolical forfeiture, are of a piece, though with potentially far more terrible consequences for the latter. While the comic absurdities that Morocco and Arragon reveal forestall or at least minimize sympathy, their plight is also a sad one, however well deserved, though they are never heard from again.

Bassanio's confidence, from first to last, is of a different quality and is motivated as much by Portia's evident favour towards him as by the belief he may have in himself (see 1.1.163–4, 3.2.1–18). Blessed with the right instincts and a good capacity to learn, when his turn comes to choose, he makes the right selection. Sceptical critics argue that Portia helps him by having a song sung as he contemplates the caskets (its first three lines all rhyme with *lead*), and it is certainly true that the other suitors are not thus privileged. The words of the song, furthermore, warn against choosing by visual attraction. However deep in contemplation he may be, Bassanio can scarcely be oblivious to it all, and his first words, 'So may the outward shows be least themselves' (3.2.73), seem to follow on directly from the song's theme. But Lawrence Danson has argued on many grounds, including the nature of the play as a romantic comedy (as opposed to a farce or 'neatly disguised satire') that the song should not be taken as Portia's 'tip-off'.[1] Furthermore, Portia's disclaimer—that she could teach Bassanio how to choose, but that would make her forsworn, something she 'will never be' (10–12)—is evidence within the play against this hypothesis. Finally, despite the ambiguities and apparent contradictions in this scene, Portia's assertion—'If you do love me, you will find me out' (41)—must carry considerable weight.[2] Again, this is part of the play's fairy-tale quality, that the handsome and noble young man, like the hero Alcides (55–7), must rescue the princess, instinctively choosing the least attractive casket precisely because, as he knows or has learned, 'The world is still deceived with ornament' (74).

Like Morocco and Arragon, Bassanio delivers a long speech before making his choice (73–107); but unlike theirs, his speech is neither filled with hyperbole (Morocco) nor swollen with abstractions (Arragon). Besides its concrete and detailed imagery, it is far more witty and perceptive.[3] The key is in his use of irony and metaphor. In contrast to Arragon's *sententiae* on 'estates, degrees, and offices', for example, and his talk of 'the true seed of honour' (2.9.41–7), Bassanio speaks of 'cowards, whose hearts are all as false | As stairs of sand' (83–4). In contrast to Morocco's confused allusion

---

[1] Danson, 118. See commentary on 3.2.63–72 and on 4.1.112 for Bassanio's apparent obtuseness in the trial scene.

[2] Cf. Nerissa's similar remarks, 1.2.27–32.

[3] Cf. a similar contrast between Romeo's lovesick laments over Rosaline (1.1.165–234) and the lines he speaks to and about Juliet (1.5.43–52, 2.2.43–67); also the contrast in the love contest between Goneril and Regan's language (*Lear* 1.1.55–61, 69–76) and Cordelia's (1.1.91–3).

to 'this shrine, this mortal breathing saint' (2.7.40),[1] Bassanio
notes how paradoxically cosmetics create 'a miracle in nature,
| Making them lightest that wear most of it' (90–1). Finally,
rejecting gold as 'hard food for Midas', he indeed gives and
hazards all he has (and more, for he risks Antonio's fortune
as well)—just as the inscription on the leaden casket demands
(2.7.9). Joy is the consequence—and ties that now bind him
to his wife.

How securely the ties are bound will be tested by the rings
which both he and Graziano—the other winner in this contest—receive from their brides with promises to wear them
always. But before the celebrations fully develop, word comes
of Antonio's misfortune. Like Mercadé in *Love's Labour's Lost*,
Salerio appears with a letter from Antonio announcing his
plight and with another couple, Lorenzo and Jessica, whose
presence recalls the raging father and creditor of 3.1.[2]
Whether, like Giannetto in Fiorentino's tale, Bassanio has
forgotten about Antonio's bond (see 'Sources', above), or
whether he has too easily acquiesced in Antonio's reassurances
(that his ships will come in before the due date, 1.3.178–9),
we cannot tell. Keeping the time scheme vague, Shakespeare
lets Bassanio remain up to now unaware of what is happening
back in Venice, though the audience is fully apprised of
events. The news therefore comes as a shock in Belmont.
Portia shows her true nobility in recognizing without the
merest hint of jealousy Bassanio's debt of loyalty to his friend
and generously offers payment many times over to deface the
bond (3.2.297–300). Afterwards, in dialogue with Lorenzo,
she comments on her motives:

> I never did repent for doing good,
> Nor shall not now; for in companions
> That do converse and waste the time together,
> Whose souls do bear an equal yoke of love,
> There must be needs a like proportion
> Of lineaments, of manners, and of spirit,
> Which makes me think that this Antonio,
> Being the bosom lover of my lord,
> Must needs be like my lord. If it be so,

[1] See commentary: shrines hold the relics of dead saints, not living ones.
[2] Leggatt, 135–6.

> How little is the cost I have bestowed
> In purchasing the semblance of my soul
> From out the state of hellish cruelty.
>
> (3.4.10–21)

Portia and Bassanio, being husband and wife, are now one (which explains why she insists on their marriage before he returns to Venice). Hence, if Antonio is as like Bassanio as he must be, given their close relationship, she feels no qualms about ransoming the 'semblance' of her soul, i.e. Antonio as a reflection of Bassanio and therefore herself. Indeed, she cannot help feeling obliged to do so, and her feelings move her to more than supplying cash. Freed from her 'enchantment' at Belmont by Bassanio's election of the right casket, she can now engage in active intervention. To do so, she must free herself and Nerissa further, from their appearance as women, so that they can be effective in the role they will play in Act 4, when the action returns to Venice.

It is a critical commonplace regarding this play that although he appears in only five scenes, Shylock dominates the action. The trial scene, in which he plays a major role, contributes to this effect, but so too does the overall dramatic structure. The casket choices, including Bassanio's, are also trials and lead up to the climactic though not final trial of 4.1. One other trial (involving the rings) takes place, helping the play to a comic denouement and further developing the kinds of bonding that make for durable and compassionate human relationships—the nature of which Shylock's role in the play helps to define.

No advanced preparation is given for his first appearance. Although Antonio has commissioned Bassanio to raise the money he needs, the only danger suggested lies in his metaphor: 'Try what my credit can in Venice do; | That shall be racked, even to the uttermost, | To furnish thee to Belmont' (1.1.180–2). By Act 4 every principal character will be 'racked' in one way or another by this action; but as 1.3 opens Shylock seems hardly anything more than a cautious moneylender, carefully contemplating the loan Bassanio desires.[1] If

---

[1] Of course, Shylock's costume may influence an audience, predisposing it accordingly as he wears a shock red wig, putty nose, and other accoutrements

Shylock enjoys keeping Bassanio on the hook as he mulls over the request, the more sinister aspects of the loan do not emerge until Shylock repeats the terms for the fourth time: 'Three thousand ducats for three months and Antonio bound' (9–10). The emphasis comes on the last word,[1] as the dialogue turns to Antonio's 'sufficiency'. Here, despite attempts at levity, Shylock lets slip a suppressed hope when he refers to the way Antonio has 'squandered' his ventures abroad (21). Slowly, but still keeping Bassanio in suspense, he suggests a positive response:

SHYLOCK Three thousand ducats; I think I may take his bond.
BASSANIO Be assured you may.
SHYLOCK I will be assured I may, and that I may be assured, I will
    bethink me. May I speak with Antonio?

(25–9)

This is Bassanio's cue, gentleman that he is, to invite Shylock to dinner, occasioning the first of two asides[2] in which Shylock reveals his deep-seated animosity to Christians generally and to Antonio in particular. It is also Antonio's cue to enter, and in his second and longer aside Shylock expresses his attitude and the underlying reasons for it:

> How like a fawning publican he looks.
> I hate him for he is a Christian,
> But more for that in low simplicity
> He lends out money gratis and brings down
> The rate of usance here with us in Venice.
> If I can catch him once upon the hip,
> I will feed fat the ancient grudge I bear him.
> He hates our sacred nation, and he rails,
> Even there where merchants most do congregate,
> On me, my bargains, and my well-won thrift,
> Which he calls interest. Cursèd be my tribe
> If I forgive him.

(38–49)

of the comic villain, or is dressed in the expensive clothes of a successful businessman. See stage history below.

[1] Danson, 140, calls attention to both its figurative and 'frighteningly literal' meanings.

[2] The first, 1.1.31–5, is not so indicated in all editions, but see commentary for its justification.

For actors and directors who wish to portray a sympathetic Shylock, the speech is problematical and therefore often cut,[1] although no one has ever questioned its textual authenticity. Retaining it, as Shakespeare doubtless intended, means endowing Shylock with motives that are as repulsive as they are comprehensible. Moreover, there is then no escaping the religious antagonism forthrightly stated in 'I hate him for he is a Christian'. But note two further points: first, his religion is not the main reason why Shylock hates Antonio; second, there is no comparable statement by any Christian in the play.[2] We have only Shylock's word that Antonio 'hates our sacred nation' and later that he despises him because he is a Jew (3.1.55). But there, too, his assertion seems qualified by Antonio's contempt for him as a money-lender. The charge of anti-Semitism in 3.1 thus comes as a non-sequitur or red herring, in consonance with this speech, where Antonio's opposition to usury carries more weight than his alleged racialism.

Not that anti-Semitism and hostility to usury can easily be separated; in fact, historically they cannot be. Although not all usurers were Jews, in drama and fiction all Jews were usurers. Under the restrictions placed on Jews throughout most of Europe, very few other occupations were open to them.[3] The origin of the twin antagonisms towards Jews and usurers may lie partly in Scripture, specifically Deuteronymy 23:20, which stipulates that interest-bearing loans may be made to 'strangers', but not to one's 'brother'.[4] As aliens, Jews were not bound in Venice or elsewhere to lend money interest-free to others outside their religious confraternity, a

[1] As in Jonathan Miller's National Theatre production with Laurence Olivier as Shylock.

[2] Although such feelings may be surmised by Antonio's comment on an 'evil soul producing holy witness' (96), or the behaviour of Solanio and Salarino toward Shylock and Tubal in 3.1, aversion, disgust, repugnance, rather than hatred, seem to characterize the Christians' attitude—until of course the trial scene, where the antagonism peaks and is most forcefully expressed by Graziano.

[3] See 'Shakespeare and Semitism', above.

[4] 'Unto a stranger thou mayest lend upon usury, but thou shalt not lend upon usury unto thy brother'; cited by John S. Coolidge, 'Law and Love in *The Merchant of Venice*', *SQ* 27 (1976), 243. Coolidge develops at length the theological aspects of the problem of usury in the play.

practice which doubtless exacerbated the antagonism engendered towards them for other reasons.

The split in the community of Venice between Christians and Jews becomes a dominant issue in 1.3 and throughout the first four acts of the play. Shylock's delayed but cordial greeting to Antonio is the first of a series of steps apparently taken to heal that breach. He follows it with the long story of Jacob and Laban, which not only emphasizes Shylock's Jewishness and his association with Old Testament patriarchs, but tries to justify his lending money out at interest. Failing at that, Shylock confronts Antonio directly, sparing few details of how the merchant prince has treated him in the past—calling him names, spitting upon his 'Jewish gaberdine', spurning him like a dog (103–25). So far from denying his behaviour, Antonio warns Shylock that he may act that way again and urges him to lend the money, if he will, not as to a friend but to an enemy. Whatever compassion he shows later in 4.1, when he and Shylock change places on the rack of judgement, he shows none now, and it remains for Shylock to try to smooth things over. Here he succeeds, saying he will forget the past, supply the money, and take 'no doit | Of usance' for it (136–7). He would be friends and have Antonio's love, he claims; and as further proof, he 'in a merry sport' will demand only a pound of flesh as the forfeit in the bond they will both sign.

His earlier suspicions and Bassanio's demurrer notwithstanding (95–9, 151–2), Antonio agrees to the bond. He recognizes Shylock's 'kindness', calls him 'gentle Jew'—very likely punning on 'gentile'—and concludes: 'The Hebrew will turn Christian: he grows kind' (174–5). At the end, the Hebrew *will* turn Christian, but out of necessity, not 'kindness'. Nevertheless, there as here 'kindness'—understood as eliminating differences ('he is our kind') as well as charity and gentleness—is the means to bind up and make whole the divided community of Venice. The question is, how sincere are the parties in this agreement? When Shylock chides Bassanio for his suspiciousness, disdaining the value of a pound of human flesh, his jocular speech convinces Antonio but leaves Bassanio (otherwise and in other circumstances so naïve) still uneasy: 'I like not fair terms and a villain's mind'

(177). Shylock may indeed be a smiling villain (such as Hamlet deplores) and will later, under provocation, drop the smiling and prove one. But Antonio is willing to take his offer as offered; and at the end, when the tables are turned in Act 4, he will follow through all the implications here of 'kindness'. Meanwhile, Shylock's attitude may be regarded as at least ambivalent.[1]

Shylock does not himself appear again until 2.5, but his presence is felt in two earlier scenes: 2.2, where Lancelot Gobbo decides to leave his master, and 2.3, where Jessica bids the clown farewell and plans to elope with Lorenzo. Her lines in this brief scene contain some harsh criticisms of Shylock but little that proves his villainy, which still remains undemonstrated. 'Our house is hell' (2) may signify nothing other than a strict, sober household in which Lancelot's antics rob it of 'tediousness'. That Jessica is ashamed to be her father's child is a more serious charge against Shylock and his 'manners', which she rejects, though it still does not prove him a villain.

In scene 5 we actually see the household members interact, and again strictness, sobriety, thrift are the watchwords. When Shylock explains why he is going out to dine with Bassanio and Antonio (contradicting his earlier demurral, 1.3.31–5), his true feelings resurface, making the 'kindness' of 1.3 seem a sham or a mere ploy:

> I am bid forth to supper, Jessica:
> There are my keys. But wherefore should I go?
> I am not bid for love. They flatter me.
> But yet I'll go in hate, to feed upon
> The prodigal Christian.

> (2.5.11–15)

[1] Danson compares Shylock here to Richard III (pp. 151–7), and Lewalski, 331 (among others) sees Shylock's attempt to heal the breach between Antonio and himself as 'merely pretence', 'a mockery of forgiveness', in so far as at the first sight of Antonio he has declared for revenge and Jessica later confirms his attitude, 3.2.282–8. But citing other parts in the dialogue, Richard A. Levin notes the ambiguity of Shylock's possibly conflicting motives: the bond may be a 'vicious and deceptive offer', or it may be an incentive to Antonio to treat him better (*Love and Society in Shakespeare's Comedy* (Newark, Del. 1985), 42). It may, of course, be both.

The resentment that Shylock feels towards the Christian community is still there. He remains sceptical of the Christian motives underlying his invitation to dinner, and whatever occurs at the gathering is no part of the play, only what happens afterwards, when Shylock suspects that the invitation was part of the elopement conspiracy. Ill at ease, with presentiments of some evil brewing, Shylock goes forth reluctantly, uncertain whether he will return immediately, giving his daughter strict orders to keep his house tightly shut against the music and merriment of masquing (28–36). 'Fast bind, fast find', he mutters—a proverb that sums up his attitude and one that Jessica, he will shortly discover, totally repudiates. Her desertion and the circumstances that surround it refuel Shylock's bitterness against Christians, destroy all feelings of 'kindness', and bind him unshakeably to a desire for revenge.

In the next scene the elopement occurs. Graziano and Salarino appear as masquers to assist the tardy Lorenzo. While they dutifully await their friend, they comment on lovers' sense of time (at the same time giving Jessica time to change into her page's costume). Graziano rather ominously expatiates on the theme, 'All things that are, | Are with more spirit chasèd than enjoyed' (12–13), lines that may suggest a subtext for Jessica, who may later have second thoughts about her defection, leaving her bemused and regretful at the end, as in the 1970 National Theatre production. Graziano concludes with the image of a ship setting off in dazzling array but returning home like a 'prodigal', with 'over-weathered ribs and raggèd sails, | Lean, rent, and beggared by the strumpet wind!' (18–19). His imagery recalls Salarino's, 1.1.8–40, attributing the cause of Antonio's melancholy to a preoccupation with his argosies 'tossing on the ocean'; more immediately, it recalls Antonio's bond, which must be paid by the safe return of at least one of those argosies. Both passages, Salarino's in the opening scene and Graziano's here, graphically describe the very real dangers of setting ventures abroad and help prepare us for the damaging news in 3.1. But as a prelude to the elopement, Graziano's speech points to the risky business the lovers are involved in as well as their prodigality, of which we also hear later in 3.1.

Jessica's nervous wit comments upon her transformation into a boy and other 'pretty follies' that lovers commit (33–42). Her actions do more. Tossing Lorenzo a casket of ducats and jewels from her balcony, she momentarily disappears to lock up the doors and 'gild' herself with additional ducats. While she is gone, Graziano remarks, 'Now, by my hood, a gentile and no Jew' (51), summing up the break from her father and his religion. Lorenzo's apostrophe to Jessica as someone 'wise, fair, and true' (53–7) conceals a discrepancy: being true to Lorenzo, she must be false to Shylock and to her religious faith. A blow 'struck at all that Shylock holds dear', her elopement is a 'crucial point in his development'[1] and leads directly to the climactic moment in 3.1 when Shylock resolves to enact his revenge against the Christian community by demanding Antonio's forfeit.

Scene 8 prepares for that moment. As Salarino and Solanio discuss Bassanio's departure for Belmont, they remark on how Shylock and the Duke wanted to search the ship for the eloped lovers. But they came too late; in any event Antonio certified that the couple was not on the ship.[2] This is the first association of Antonio with Shylock's personal loss, and the first time Shylock is explicitly called 'villain' (2.8.4). Solanio laughs at the old man's distress at losing his daughter and his ducats (12–22), but ends with the warning: 'Let good Antonio look he keep his day, | Or he shall pay for this' (25–6). Earlier hints that the merchant's ships may founder on the seas are now more concrete (27–32); however, in his emotional leave-taking Antonio urges Bassanio to banish all thoughts of the bond: 'And for the Jew's bond which he hath of me, | Let it not enter in your mind of love: | Be merry . . .' (41–3). His friend's departure, on the other hand, leaves Antonio far from merry; rather, his melancholy—'his embracèd heaviness' as Solanio calls it—tightens its grip, and worse will follow.

The alternation of good and bad luck—Morocco and Arragon failing in their quest, Jessica and Lorenzo successfully

---

[1] Midgley, *'Merchant of Venice'*, 124.

[2] Divided as the Venetian community may be, the Christian and Jewish segments are themselves closely knit; hence, Shylock suspects that Bassanio is a party to Jessica's elopement, and in 3.1 he further associates Solanio and Salarino and then Antonio with it.

eloping, Shylock losing his daughter and his ducats, Lancelot winning a new master—culminates in Act 3. In Belmont, immediately Bassanio wins Portia by choosing the right casket, word arrives that Antonio has certainly lost his ships and is in danger of surrendering his forfeit. Before that, a climactic scene of another sort occurs, where fear and jocularity also alternate and end with threats of dire disaster. Just as his characters experience joy and sadness almost simultaneously, Shakespeare encourages his audience to entertain similar feelings, thereby intensifying the comic pleasure that comes at the end when everything—or almost everything—is resolved in a relatively happy outcome.

The intermingling of joy and fear in Act 3 begins with Solanio and Salarino's dialogue, lamenting the loss of Antonio's ship on the Goodwins, and fearing it is not the last of his losses. At this moment Shylock enters in disarray, bitterly accusing them of conspiracy in his daughter's elopement. Relieving their anxiety over Antonio, Solanio and Salarino mock Shylock until Salarino's anxiety gets the better of him and he asks Shylock if he, too, has heard of Antonio's misfortune. At first, the remark only deepens Shylock's sense of loss, but almost at once he sees an advantage:

There I have another bad match. A bankrupt, a prodigal, who dare scarce show his head on the Rialto; a beggar, that was used to come so smug upon the mart. Let him look to his bond. He was wont to call me usurer: let him look to his bond. He was wont to lend money for a Christian courtesy: let him look to his bond. (41–7)

The thrice uttered 'Let him look to his bond' alerts Salarino to the danger, which he tries to laugh off, nervously asking 'What's that good for?' In reply, Shylock first is mordantly flippant—'To bait fish withal'—then ominous: 'If it will feed nothing else, it will feed my revenge' (50–1). To justify taking revenge, he launches into a long speech affirming his identity as a Jew but also as a human being—no better and no worse than others are, including Christians (55–69). And it is Christian example, he claims, that has taught him the nature of revenge.[1] Bound to the rest of humanity by identical attributes

---

[1] Here Shakespeare carefully distinguishes between Christian practice and Christian doctrine.

of sense and feeling as well as a shared mortality, he feels bound also by a common dedication to revenge for wrongs suffered at the hands of his enemy. *This* 'kindness' he *will* show (1.3.140).

Out of context, without the framework of vengeance, the speech is an eloquent plea for Jews—and by extension, other minority groups—to be viewed as human beings also; it has often been quoted or recited as such. An emotional high point in the play, it easily earns 'tragic' Shylocks enormous compassion. But even for less attractive Shylocks the speech also arouses sympathy, even though it appeals to the lowest common denominators among human attributes, as many among Shakespeare's audience would notice. By endowing his villainous money-lender with 'organs, dimensions, senses, affections, passions', Shakespeare identifies Shylock with the rest of humanity, but a humanity depraved by revenge. Utterly absent from Shylock's claims is any connection with purer or nobler human attributes. Furthermore, he fails to consider here or elsewhere that neither his religion nor Christianity condones or sanctions vengeance. Quite the contrary: both systems of religious belief stand adamantly opposed to it, condemning it as a corruption of the human spirit.[1]

In Act 4 Portia's plea to be merciful is designed to help Shylock overcome his human depravity and redeem himself in a way he finds impossible to understand or pursue. Here, the conflicting emotions aroused in him by the loss of his daughter and his money, the taunting of Salarino and Solanio, and Tubal's news alternating between reports of Antonio's misfortune and Jessica's spendthrift ways evoke in the audience a complex response mingling laughter, repugnance, and compassion. We sense the wrongs Shylock suffers and the agony he feels, for example, when Tubal describes how Jessica gave his turquoise 'for a monkey'—the ring his wife Leah had given him when he was a bachelor, though his hysteria is

---

[1] See e.g. Lev. 19: 17–18; Ps. 94; Rom. 12: 19. In the sixteenth century, both church and secular authorities vigorously opposed blood vengeance. Compare John R. Cooper, who says that Shylock's speech provides insight into his 'all-too-human mind. We see his motives arise from a nature like ours, one that looks for vengeance when wronged.' Nevertheless, 'Shylock, in seeking revenge against Antonio, is not justified' ('Shylock's Humanity', *SQ* 21 (1970), 120).

also comical (111–16). And revulsion comes as Shylock exults over Antonio's losses—'I'll plague him, I'll torture him. I am glad of it' (109–10). Finally, like the Duke (4.1.16–20), we cannot believe he will really exact the grotesque forfeiture the bond requires.

His heart hardens against Antonio in the brief scene after Bassanio's success at Belmont. Shylock specifically rejects any suggestion of mercy (3.3.1) and derides all 'Christian inter-cessors' (14–16), thus anticipating the opposition of justice and mercy in the trial scene. In vain, Antonio begs to be heard, but Shylock will not hear; he will have his bond, he repeats over and over again. It is a curious scene—Antonio, led by a gaoler, pleading with Shylock to hear him; Shylock, chiding the gaoler and waving Antonio off, insisting 'I'll have no speaking; I will have my bond' (17). He clearly has the upper hand, as Antonio, recovering his composure and his dignity, dashes Solanio's hope that the forfeit will not hold: 'The Duke cannot deny the course of law', he says, for if the penalty is denied, it will 'much impeach the justice of the state' (26–9). But in explaining Shylock's motivation, he mentions only financial rivalry, nothing about religious dif-ferences:

> He seeks my life. His reason well I know:
> I oft delivered from his forfeitures
> Many that have at times made moan to me;
> Therefore he hates me. (21–4)

Their conflict over the competing claims of justice and mercy has often emerged before, it seems, but never in such deadly earnest as now.

As an allegory of Justice and Mercy, the Old Law versus the New, is the way some critics interpret the trial scene, 4.1.[1] Doubtless some allegorical aspects are present. By his own words, Shylock 'stands for' judgement and for law (102, 141)—for 'justice', narrowly conceived. Identified in the scene immediately preceding as an embodiment of heavenly joys here on earth (3.5.71), Portia here speaks on behalf of divine

---

[1] See e.g. Nevill Coghill, 'The Basis of Shakespearian Comedy', *Essays and Studies*, NS 3 (1950), 20–3; C. L. Barber, *Shakespeare's Festive Comedy* (Prince-ton, NJ, 1959), 185; and Lewalski, 331–5.

mercy (4.1.181–200). But of course the trial is more than a medieval allegory, just as Shakespeare's drama is more than a morality play, though it incorporates elements of both. Portia as Balthasar, dressed in her robes, is very much the young legal expert, 'brisk and businesslike';[1] Shylock is very much the wronged individual and a grimly determined creditor, demanding his due. Neither embodies merely an abstract concept as an allegorical figure. What makes the scene defy ordinary reality (besides the fact that it is all play-acting) are the dynamics of the conflict and the extraordinary circumstances. What began as a 'merry sport' (ostensibly to heal up the breach between money-lender and merchant, Jew and Christian) has become deadly serious.

In demanding his forfeit, Shylock, however, seems to lose most of his humanity—not as he defines it, but as others do. Epithets hurled at him earlier now take on added significance. More than a 'stony adversary', he is 'an inhuman wretch | Uncapable of pity' (3–4).[2] Antonio, almost Christ-like, opposes his 'patience' to Shylock's fury and is 'armed | To suffer with a quietness of spirit | The very tyranny and rage of his' (10–12).[3] The first appeal for mercy comes from the Duke (16–33), whose words suggest (though they never explicitly state) that a positive response would help heal the breach in Venice, especially as he asks Shylock to be generous and forgive both the forfeit and part of the principal as well. But Shylock is adamant: he has taken an oath 'by our holy Sabbath' to have 'the due and forfeit' of his bond (35–6). Oblivious to the oath's blasphemy, he argues the consequences to the city if his claim is denied. Bassanio intervenes, but Antonio sees the futility of argument and for the first time cites Shylock's religion as a source of cruelty: nothing is harder than his 'Jewish heart' (79). Hoping greed will overcome cruelty, Bassanio offers to double the principal, all in vain. Now the Duke intervenes: 'How shalt thou hope for mercy, rend'ring none?' (87). Like a Pharisee maintaining

---

[1] Granville-Barker, 88.

[2] Graziano, citing Pythagoras, suggests that the spirit of a wolf, hanged for human slaughter, 'infused itself' in Shylock while he was still an embryo (131–8).

[3] For Antonio as 'a passive, submissive Christ, risking his body and blood,' see Holland, *Psychoanalysis and Shakespeare*, 331.

his righteousness, Shylock defies judgement. He is convinced, moreover, that the law is on his side. To bring home his point, at the same time exposing Christian hypocrisy, he compares their ownership and treatment of slaves to his claim to own a piece of Antonio (89–100).

The Duke recognizes the impasse, which he hopes Dr Bellario can resolve. But instead of the learned jurist, young Balthasar appears. Meanwhile, Bassanio offers to trade places with Antonio who, resigned to self-immolation, sees himself as a kind of sacrificial victim, 'a tainted wether of the flock, | Meetest for death' (113–14).[1] Bassanio's offer points to the eventual resolution of the problem; but Shylock, whetting his knife, on his part also intent on Antonio's demise, misses the clue, just as Bassanio and the others do:[2]

> Good cheer, Antonio! What, man, courage yet!
> The Jew shall have my flesh, blood, bones, and all
> Ere thou shalt lose for me one drop of blood.
>
> (110–12)

After a further interlude (an acrimonious exchange between Graziano and Shylock), the focus centres on Bellario's emissary and the trial, although technically it is not that but a hearing.[3] And it is here that the conflict between Justice and Mercy reaches its apex.

[1] From this point onwards, the scene suggests an analogy also to the Sacrifice of Isaac, in which Isaac is spared at the last moment, and a ram is substituted for him (Gen. 22). In such analogues, parallels are seldom exact, though they resemble each other in general shape and movement, and more than one analogue may operate simultaneously. For example, Antonio as the intended sacrificial victim may parallel Isaac, notwithstanding the reference to himself as a 'wether' (= ram), i.e. the substitute. On his role as a scapegoat or 'sin offering', atoning for the sins of the community, compare Lev. 16 : 5–16. For Antonio as the victim of 'father aggression' in psychoanalytic terms, see Holland, *Psychoanalysis and Shakespeare*, 235–7.

[2] One reason for this apparent obtuseness may be that everyone implicitly recognizes the legal principle that 'a right granted to do something includes the right to do anything that is necessarily incidental thereto', as the nineteenth-century German jurist, Josef Kohler, maintained. Moreover, Portia's second argument is vitiated in so far as a creditor is allowed to take less than that to which he is entitled. But these points do not supersede the more important one involving an abuse of legal right: 'to seek to take a person's life through abuse of right is against law as attempted murder'. See O. Hood Phillips, *Shakespeare and the Lawyers* (1972), 95, and Merchant, 24.

[3] For many opinions concerning the nature and conduct of the trial, see e.g. Phillips, *Shakespeare and the Lawyers*, 91–118, and compare Tucker, 'The

Portia begins by recognizing Shylock's claims under Venetian law and, gaining Antonio's concurrence, concludes: 'Then must the Jew be merciful' (179). When Shylock bridles at her suggested 'compulsion', she launches into a speech on the nature of mercy and its benefits to both giver and receiver: 'It is twice blest', she says: 'It blesseth him that gives and him that takes' (183–4). She then expands upon the Duke's earlier theme:

> Therefore, Jew,
> Though justice be thy plea, consider this:
> That in the course of justice none of us
> Should see salvation. We do pray for mercy,
> And that same prayer doth teach us all to render
> The deeds of mercy.
>
> (194–9)

But Shylock cares nothing for 'deeds of mercy'. Echoing those who demanded Jesus's crucifixion, he cries: 'My deeds upon my head! I crave the law, | The penalty and forfeit of my bond' (203–4).[1] The impasse remains as it was.

Further offers to repay the debt several times over avail nothing. Nor do Portia's repeated appeals for mercy (230, 254–5, 258). But by also affirming the claims of justice under the law—and the bond's apparently airtight claim—Portia does more than win Shylock's extravagant praise ('A Daniel come to judgement, yea, a Daniel!'). She establishes the context in which mercy gains its full significance as a fulfilment of law, not its antagonist.[2] Or, in less theological terms, she demonstrates how mercy becomes meaningful only in the context of justice. Failing first to recognize the claims of justice, we may fall too easily into forgiveness. To avoid such

Letter of the Law', 93–101. Among some of the points discussed are the validity of the bond in the first instance, Portia's role as an interested party, and the differences between common law and equity. Lawyers have a field day with this scene, but legal technicalities are mostly irrelevant to the scene as Shakespeare presents it, which carries sufficient plausibility for his *dramatic* purpose.

[1] See commentary on the reference to Matt. 27: 25.

[2] Compare Danson, 65. Citing St Augustine's commentaries, he says that 'the essential thing added to the law by Christ is forgiveness. Mercy, therefore, is made part of the law, rather than an opposing principle. Indeed, mercy, or forgiveness, becomes the legal principle enabling all other legal principles'.

sentimentality, Shakespeare clarifies what justice involves before moving to the administration of mercy, just as he does
later in *Measure for Measure*.[1] At the same time, he gives
Shylock every opportunity to see this for himself. But as E.
M. W. Tillyard maintains, Shylock is the victim of 'spiritual
stupidity';[2] he cannot see the moral advantage in rendering
mercy any more than he can perceive the moral danger
inherent in revenge. When Portia stops him just as he thinks
he has succeeded in his quest, he becomes hoist by his own
petard, as befits his insistence on 'justice', or what Danson
calls his 'diabolical literalism' (p. 96).

By rejecting Portia's several invitations to show compassion,[3] Shylock fails to rise above the lowest level of humanity
and convert incipient tragedy into the 'merry sport' he
claimed it was when he first proposed the bond (1.3.134–48).
Therefore, Portia must resort to other measures, in the process
assuring a comic outcome.[4] Not only must she demonstrate
to Shylock the error of a fanatical dependence upon literalist
judgements, she must further humanize him so that he
becomes acceptable within the community. Although for some
critics her first procedure depends upon a legal quibble, for
others it is not merely a verbal trick. It is, rather, 'the
revelation of truth', which in part consists in showing that
the law is 'society's servant, not its master, and . . . the
unmitigated law before all who would stand condemned can
be made to yield its mercies'.[5]

Shylock has asked for 'justice' and receives more than he
desired under the same terms and conditions he has insisted
upon. Taking his argument to its logical extreme, Portia
awards him his just pound of flesh but not one drop of blood.
It is now she who appears to be as relentless as Shylock was

[1] Compare Frank Kermode, 'The Mature Comedies', in John Russell Brown
and Bernard Harris (eds.), *Early Shakespeare* (1961), 223: 'as Shakespeare is
careful to show in *Measure for Measure* the arguments for justice are strong,
and in the course of Christian doctrine it is necessarily satisfied before mercy
operates'.

[2] *Shakespeare's Early Comedies* (1965), 192.

[3] On her graduated appeals to Shylock to 'prove his affinity with humankind', see Bertrand Evans, *Shakespeare's Comedies* (Oxford, 1960), 64.

[4] See Barber, 180–5.

[5] Danson, 119–120.

earlier. He shall have 'all justice', she insists; hence, she now disallows the offer of thrice the principal or even the principal itself. She awards Shylock nothing but the penalty—under the further condition that the pound of flesh exceed the just weight by not so much as 'the twentieth part | Of one poor scruple' (320–8). Paradoxically, she frees Antonio from the tyranny of the bond by focusing more strictly than Shylock on the binding quality of its language. Defeated, Shylock starts to leave, when Portia begins her second attack.

Convicting him of violating another law of Venice, Portia now holds Shylock condemned to death and all his goods confiscate, half to the state and half to the party upon whose life he had practised. It is here, in the strictest context of justice, that her eloquent speech on mercy bears fruit. Graziano's incorrigible vindictiveness (360–3) notwithstanding, the Duke at once spares Shylock's life. Further, he suggests that Shylock's humility can drive that part of his fortune owed the state into a fine. But Shylock is not yet humbled; a burst of the old fire remains:

> Nay, take my life and all! Pardon not that!
> You take my house when you do take the prop
> That doth sustain my house; you take my life
> When you do take the means whereby I live.
> (370–3)

It is now Antonio's turn to demonstrate what he has learned from Portia's speech. With little prompting, he asks the Duke to remit the fine to the state for half of Shylock's goods, thereby restoring the 'prop' that will sustain his 'house'. He specifies two further conditions: first, that Shylock become a Christian; second, that he make Lorenzo and Jessica his sole heirs.

Antonio's first condition perhaps more than anything else in the play troubles modern audiences.[1] Regardless of how Elizabethans responded, audiences today find the imposed conversion difficult to accept. Reinforced by the Duke's threat

---

[1] Compare Sinsheimer, *Shylock*, 99: 'That Shakespeare makes Shylock agree to be baptised is the worst offence of all'; Nevo, *Comic Transformations*, 136–7: 'The benign offer of [conversion] is the ultimate cruelty of alienation, of denial of that essential being which has just made itself so palpably manifest. Therefore it is counterfeit mercy.'

to rescind his pardon, Antonio's requirement may be essentially 'merciful'—Shylock's only opportunity to win eternal life as well as his mortal one.[1] But in an age more secular than Shakespeare's, the requirement appears to violate human rights. However intended to complete the comic structure of the play, it remains for a modern audience problematical. Seen from a social rather than theological perspective—the attempt at last to purge Shylock's humanity and welcome him more fully into the community as 'one of us'[2]—Antonio's demand and Shylock's acquiescence are still seriously disturbing. That is one reason why *The Merchant of Venice* appears to some more a 'problem play' than a 'festive' comedy'[3] and its fifth act more like an excrescence than an appropriately celebratory ending.

Part of the problem lies in Shylock's ready willingness to accept Antonio's terms. A more deeply religious person would prefer death to apostasy—so much for Shylock's pretended devotion to his faith![4] From his apparently swift compliance, some actors playing the role regard Shylock as a man finally more interested in money than anything else, including his daughter and his religion.[5] But this is to make too summary a judgement, oversimplifying Shylock's character and the shadings latent in his departing lines. As Portia puts the question squarely before him, Shylock hesitates before answering:

> PORTIA
> Art thou contented, Jew? What dost thou say?
> SHYLOCK
> I am content.

(389–90)

---

[1] See Coghill, 'Basis of Shakespearian Comedy', 22–3; Lewalski, 341; Grebanier, *Truth about Shylock*, 291; Kirschbaum, 'Shylock and the City of God', 29–31.

[2] '... perhaps Elizabethans saw in [Shylock's conversion] a gesture in the direction of bringing him into the community of civilized men', 'Introduction', Sylvan Barnet (ed.), *Twentieth Century Interpretations of 'The Merchant of Venice'* (Englewood Cliffs, NJ, 1970), 7.

[3] Compare W. H. Auden, 'Brothers & Others', *The Dyer's Hand* (New York, 1962), 223–5, and Leo Salingar, 'Is *The Merchant of Venice* a Problem Play?', in *Dramatic Form in Shakespeare and the Jacobeans* (Cambridge, 1986), 19–31.

[4] If life is at risk, Jews may violate any of the 613 commandments in the Old Testament—except this one, which can require martyrdom.

[5] e.g. Patrick Stewart (see stage history, below).

His few words of compliance lend themselves to various nuances—deep reluctance, weary acceptance of the inevitable, bitter resignation, moral collapse, relief that he has escaped the worst and has a chance to live despite reduced resources, and so forth. They are followed by lines that also signal distress, as Portia orders Nerissa to prepare the deed of gift:

> I pray you, give me leave to go from hence.
> I am not well. Send the deed after me,
> And I will sign it.
>
> (391–3)

Shylock's last words in the play—monosyllabic and fraught with heaviness—may be spoken with some eagerness to be gone, to withdraw from the sight of his tormentors; but they may also convey the weight of his defeat simultaneously with a new understanding of shared humanity. The diabolical lust for revenge—or the Old Adam, as some critics see it—has been whipped out of him, and he is left exhausted, no longer a serious menace or a comic butt. He is human—like the rest of the community, if not yet quite fully part of it, as the Duke's gruff dismissal and Graziano's parting taunts suggest. However he was played by Shakespeare's colleague, the actor who can communicate to a modern audience as many conflicting emotions as possible does best justice to the role.

It remains only for the other principals to recover their true identities, return home, and become reunited in joyful harmony. To avoid sentimentality, extend the trial motif in a lighter comic vein, and conclude the significance of bonds and bondage, Shakespeare includes the ring plot and adds some further discordances, mainly in the rather incongruous figure of Antonio at Belmont. Other jarring notes occur in the teasing banter between Lorenzo and Jessica. Lorenzo opens the fifth act with some lovely lines about the moonlight and the gentle breezes; but then in what is at one level mere playfulness, he and Jessica begin a series of allusions to unhappy love stories involving Troilus and Cressida, Pyramus and Thisbe, Dido and Aeneas, Medea and Jason—among which Lorenzo humorously associates his elopement with Jessica (5.1.3–20). Before the scene shifts to Belmont, however, Portia and Nerissa concoct their plot to try the fidelity of their husbands.

The ring Portia gives to Bassanio represents, she says, all that she has and is; and when he parts from it, by losing or giving it away, it will presage the ruin of his love (3.2.171–4). Bassanio responds accordingly, solemnly swearing to keep the ring on his finger until death. His predicament is therefore acute when, declining monetary recompense, Portia as Balthasar asks for the ring. Only Antonio's intervention persuades Bassanio to agree, bringing to bear the conflict of Bassanio's two loves.[1] Bound to Portia by his marriage vows, he is bound also to Antonio by friendship and even deeper obligations: this is the man whose flesh, his very life, was pawned for him. He yields the ring after Antonio puts the issue directly:

> Let his deservings and my love withal
> Be valued 'gainst your wife's commandement.
>
> (4.1.446–7)

Of course, Portia means to get the ring back as a joke, as she says to Nerissa, who in turn gets hers from Graziano (4.2.13–17). But jokes, as Freud long ago showed, often convey more than humour, or rather the humour overlays a deeper, sometimes tendentious, significance. Earlier in the trial scene, when things looked especially grim, Bassanio had passionately exclaimed that he would sacrifice everything— life, wife, and 'all the world'—to deliver Antonio from Shylock's grip (4.1.279–84). He is seconded by Graziano, who also swears he would willingly sacrifice his wife to free Antonio. These outbursts occasion comic asides from Portia and Nerissa, as well as a bitter one from Shylock ('These be the Christian husbands', 4.1.292), but the men's outbursts evidently sink in, and the women are out to teach their husbands a lesson.

The weapon they use to bring home their point about the supremacy of the marriage contract is the threat of infidelity.

[1] On the development of this conflict, see Coppélia Kahn, 'The Cuckoo's Note: Male Friendship and Cuckoldry in *The Merchant of Venice*', in Peter Erickson and Coppélia Kahn (eds.), *Shakespeare's 'Rough Magic'* (Newark, Del., 1985), 104–12.

They eschew any allegiance to Elizabethan double standards and assert their sexual equality. Since their husbands have been unfaithful, they will be too—with the very ones to whom they gave the rings. Revenge, so much the subject of the earlier acts, rears its head again here.[1] Although the audience is in on the joke, Bassanio's anguish and Graziano's discomfiture show that the lesson is working. When Antonio intervenes, recognizing that he is the 'unhappy subject of these quarrels' (5.1.238), he again pledges himself on Bassanio's behalf, but this time with his soul, not his flesh, and the conflict is resolved.[2]

The broken harmonies with which the scene began are now wound up. Portia, who presided over the trial in Act 4, here presides over the dropping of 'manna in the way | Of starvèd people' (294–5). Conventionally, as Harry Levin reminds us, the happy endings of comedies are formalized by various forms of revelry—feasts or dances—with a betrothal or mating in view.[3] But since the weddings have already taken place, it is time now for the couples to retire to the bridal chamber, and for other reasons revelry here might seem inappropriate. For one thing, the ending perforce excludes many characters, chiefly Shylock.[4] For another, Antonio (unlike his counterpart in Shakespeare's source) remains a solitary figure—however his isolation may be disguised—and thus disturbing.[5]

[1] Kahn, 'The Cuckoo's Note', 108–11, develops these points in detail.

[2] Compare Leslie Fiedler, *The Stranger in Shakespeare* (New York, 1972), 135, who says that in giving him the ring to give to Bassanio, Portia takes her revenge on Antonio, by forcing him to marry her to Bassanio a second time, as it were.

[3] 'A Garden in Belmont: *The Merchant of Venice*, 5.1', in W. R. Elton and William B. Long (eds.), *Shakespeare and Dramatic Traditions* (Newark, Del., 1989), 29.

[4] Leggatt, 149: 'In short, the play has shown a larger world than it can finally bring into harmony.'

[5] Compare Auden, 'Brothers & Others', 233–4: 'If Antonio is not to fade away into a nonentity, then the married couples must enter the lighted house and leave Antonio standing alone on the darkened stage, outside the Eden from which, not by the choice of others, but by his own nature, he is excluded'. See also Leonard Tennenhouse, 'The Counterfeit Order of *The Merchant of Venice*', in Murray M. Schwartz and Coppélia Kahn (eds.), *Representing Shakespeare* (Baltimore, 1980), 63; and Jean Howard, 'The Difficulties of Closure', in A. R. Braunmuller and J. C. Bulman (eds.), *Comedy from Shakespeare to Sheridan* (Newark, Del., 1986), 125, who argues that in the end the ironies not the harmonies are the most striking feature of this play.

Although Lorenzo's melodic verses describe the music of the spheres, he recognizes that 'we cannot hear it' (54–65). Music surrounds the lovers, and for some the image of Jessica and Lorenzo in each other's arms (109) is an emblem of Christian and Jew, the New Law and the Old, united in love.[1] But for others the tone of the ending has much of the ironic reserve of Christ's praise of the bad steward (Luke 16:8).[2]

We end where we began, acknowledging the play's inconsistencies and contradictions, but at the same time its plenitude, its richness—a richness so great that the foregoing discussion has necessarily omitted or slighted a number of important aspects. More could be said of love's wealth as opposed to crass materialism, as in John Russell Brown's analysis[3] and in C. L. Barber's, which opposes Antonio's loan (viewed as 'venture capital') to usury, or money used to get money.[4] Something, too, could be said about the way the play reflects and perhaps comments upon the development of capitalism at the end of the sixteenth century—the conflict between Antonio's modern capitalist values (from an Italian point of view) and Shylock's quasi-feudal fiscalism, and indeed the opposition of various class perspectives.[5] M. M. Mahood is doubtless right in refusing to yield to critical reductiveness, arguing that such underlying unity and coherence as the play has remains intuitive, hence within the individual possession of each member of an audience.[6] The broad approach taken here—showing how concern with bonds and bondage pervades the play—is not all-encompassing. Criticism cannot displace art, after all; the play's the thing, as it always was, and is. And as the stage history of *The Merchant of Venice*

[1] Coghill, 'Basis of Shakespearian Comedy', 23. Levin, 'A Garden in Belmont', 24, notes that music, the recurrent symbol of harmony in Shakespeare, as Coghill says, is alluded to fifteen times, more than in any other Shakespeare play, and eleven times in the last act alone.

[2] A. D. Moody, *The Merchant of Venice* (1964), in Barnet, *Twentieth Century Interpretations*, 107.

[3] 'Love's Wealth and the Judgement of *The Merchant of Venice*', in *Shakespeare and his Comedies* (1957), 62–75.

[4] 'The Merchants and the Jew of Venice: Wealth's Communion and an Intruder', in *Shakespeare's Festive Comedy*, 163–91, esp. 175.

[5] See Walter Cohen, '*The Merchant of Venice* and the Possibilities of Historical Criticism', *ELH* 49 (1982), 765–89.

[6] NCS 25.

demonstrates, the play lends itself to a wide variety of interpretations—so various that no one staging can encompass the
full range available. But, taken together, they may comprise
a comprehensive criticism that makes us aware of the many
currents and counter-currents that flow within *The Merchant
of Venice*.

The Merchant of Venice *in Performance*

Over the centuries, *The Merchant of Venice* has been one of
the most frequently performed of Shakespeare's plays in the
professional theatre. The reasons for its popularity among
producers, actors, and audiences are not hard to infer. It is
a play of rich complexity, raising issues of justice and mercy
(among others) that are not easily resolved—if resolved at all.
It is also a great starring vehicle, pre-eminently in the roles
of Shylock and Portia but also with excellent character parts
in Bassanio, Graziano, Nerissa, Jessica, Lorenzo, and Lancelot
Gobbo. The double-plot structure involving Venice and Belmont allows the set and costume designer wide scope for
inventive splendour. A romantic comedy that borders on, and
in some ways penetrates, the environs of tragedy, *The Merchant* has captivated audiences for over two centuries, or ever
since Granville's debasement, *The Jew of Venice* (1701), was
replaced by something closer to Shakespeare's original.

For Shakespeare's play as presented in this edition was by
no means always identical with the acting script used in the
theatre. More often than not until recently—and even now
quite often—the text underwent revisions of one kind or
another involving omissions (sometimes of an entire act),
additions, transpositions of passages or scenes—anything, in
short, that in the view of the director or producer might make
for a good 'show'. This fact will surprise no one at all familiar
with the inner workings of the theatre, whose goal, in Shakespeare's time as in ours, has always been to ensure good
box office receipts. The purist, of course, may long for 'authentic' Shakespeare. But that desideratum remains a will-o'-
the-wisp, a working and workable definition of which has
eluded scholars and annoyed theatre practitioners, intent as
they are on something more vital and alive than 'museum'

Shakespeare.[1] Meanwhile, we have a robust series of productions and performances that endlessly explore, develop, and present the rewarding depths of Shakespeare's art.

Seventeenth-century productions

Unfortunately, scant information on *The Merchant of Venice* as performed in Shakespeare's era has come down to us. The first known record of any performance appears on the title-page of the first quarto (1600; see 'Textual Introduction', below). There we learn that the play was performed 'divers times' by the Lord Chamberlain's Men. No date is given for any performance, but probably the play was in the repertory by 1596-7.[2] The next record is of performance at the court of James I on Shrove Sunday (10 February) 1605. James apparently liked the play, since he ordered it to be repeated at court two days later on Shrove Tuesday. On the Monday between, the King's Men (as they were now called) performed *The Spanish Maze*, a play about which nothing else is known.[3]

The appropriateness of *The Merchant of Venice* to be performed at Shrovetide, like its more recent popularity, may not be difficult to apprehend.[4] The season is one that mixes carnival licence and penitential introspection with the desire, indeed the need, for absolution. Shrove Tuesday ('Fat Tuesday') is the day when carnival reaches its apex, the day before Ash Wednesday and the beginning of Lent. Some have suggested that *The Merchant of Venice* is in fact set during Shrovetide and the Lenten month that follows.[5] Certainly the masquing discussed by Lorenzo, Graziano, and the others in Act 2 suggests the period of carnival, though in fact the masquing does not take place, and in any case masquing is not limited to Shrovetide. More pertinent, perhaps, are the thematic suggestions of ritual sacrifice, God's mercy, and the

[1] For a discussion of this problem, see Peter Brook, 'The Deadly Theatre', ch. 1 of *The Empty Space* (New York, 1968; repr. 1978), 9-41.

[2] See 'Sources, analogues, and date' above. Cf. E. K. Chambers, *The Elizabethan Stage* (Oxford, 1923), ii. 195, and *TC* 119-20.

[3] Ibid. iv. 119, 172.

[4] For several of the ideas here I am indebted to R. Chris Hassel, *Renaissance Drama and the English Church Year* (Lincoln, Neb., 1979), 113-18.

[5] Hassel, *English Church Year*, 118, who cites Enid Welsford's descriptions of traditional Shrovetide public masquing, mummery, etc., in *The Court Masque*, 2nd edn. (1927; repr., New York, 1962), 12, 36.

grace of love that dominate both the play and the liturgy of Shrovetide and that would appeal to the theologically oriented James.

Whatever the situation at court, most likely in the first performances Will Kemp played the part of the clown, Lancelot Gobbo, which is typical of his presumed roles in other early Shakespearian comedies. On the other hand, if Shylock was played as a comic villain, he rather than Burbage may have essayed the role.[1] Doubling was, of course, a common practice among Elizabethan and Jacobean acting companies, which depended upon a nucleus of about a dozen actors and apprentices plus a number of mutes, or supers, as we should now call them, to stage their plays. According to one analysis, twelve men and four boys, or sixteen actors (including mutes), were required for *The Merchant of Venice*, which contains twenty speaking parts and eight or nine distinguishable mutes needed, for example, to fill up Morocco's train and Portia's in 2.1 and 2.7. This is the usual number of actors Shakespeare depended upon at this time, that is, before moving into the Globe Theatre a few years later, although *Richard III*, with fifty-five speakers and seventeen mutes, required as many as seventeen actors.[2]

Following the performances at court in 1605, no others are recorded until after the Restoration, when *The Merchant of Venice* was among those of Shakespeare's dramas assigned to the actor-manager, Thomas Killigrew.[3] When the play dropped out of the King's Men's repertory—or if it did—is not known, nor is any record of a performance by Killigrew's company. Very likely the play, like other comedies by Shakespeare, had little appeal to the more refined tastes of the later Stuart period, influenced as it was by French drama, which now set the standard for much of English culture. It seems

[1] Toby Lelyveld, *Shylock on the Stage* (Cleveland, 1960), 7.

[2] William Ringler, Jr., 'The Number of Actors in Shakespeare's Early Plays', in G. E. Bentley (ed.), *The Seventeenth-Century Stage* (Chicago, 1968), 123.

[3] The play is included along with fifteen others by Shakespeare and many more plays by Jonson, Beaumont and Fletcher, Massinger, etc., formerly performed at the Blackfriars Theatre and now assigned to Thomas Killigrew and the King's Company on 12 Jan. 1669. (The rest were assigned to Davenant.) See *The London Stage, 1660–1800*, ed. William Van Lennep *et al.* (Carbondale, Ill., 1965), 5 vols. in 11; Part I, 1660–1700, 151–2.

inevitable, therefore, that *The Merchant of Venice* would eventually experience the fate of *The Tempest* and *Troilus and Cressida*, as well as tragedies like *King Lear* and *Macbeth*, which underwent rewriting and 'improvement' at the hands of lesser but more fashionable playwrights, in this instance George Granville, who rewrote it in 1701.

The Merchant of Venice *transformed*

Granville's adaptation, called *The Jew of Venice*,[1] is typical of the drama of its time, particularly of Shakespearian transformations. Like Tate's *King Lear* (1681), *The Jew of Venice* preserves many of Shakespeare's lines but adds others; omits characters; shortens, transposes, and cuts passages or scenes; and invents a remarkable new scene—an elaborate banquet at the end of the otherwise heavily curtailed Act 2. Shylock is present, and all the guests witness an elegant masque entitled 'Peleus & Thetis'. Gone, however, are many comic characters, not only Lancelot Gobbo and his father, but Morocco and Arragon. Gone too are Salanio, Salerio, and Salarino; Tubal; Stefano; and other minor characters. Granville attempted thereby to condense the play, partly to make room for the masque, but apparently also to adhere more closely to neo-classical concepts of dramatic unity.

Two examples of Granville's adaptation illustrate both his method and the taste of his time. The play opens, not with Salanio and Salarino probing the cause of Antonio's melancholy, but more than seventy-five lines later (Granville inserts inverted commas in the text to indicate major additions or revisions, though not all of them):

> *Anto.* I Hold the World, but as a Stage, *Gratiano*,
>   'Where every Man must play some certain Part,
>   And mine's a serious one.
> *Grat.* Laughter and Mirth be mine,
>   Why should a Man, whose Blood is warm and young,

[1] Granville's name does not appear on the title-page of the quarto, which advertises the play as 'Acted at the Theatre in Little-Lincolns-Inn-Fields, by His Majesty's Servants' and published by 'Ber. Lintott at the Post-House in the Middle Temple-Gate, Fleetstreet, 1701'. It is reprinted in *Five Restoration Adaptations of Shakespeare*, ed. Christopher Spencer (Urbana, Ill., 1965), 345–402, and reproduced in facsimile by Cornmarket Press (London, 1969). References and quotations below are from the facsimile.

> Sit like his Grandsire, cut in Alablaster! [*sic*]
> Sleep, when he wakes, and creep into the Jaundice,
> By being peevish! I tell thee what, *Antonio*!
> I love thee, and it is my Love that speaks;
> There are a sort of Men, whose Visages
> Do cream and mantle, like a standing Pond;
> And do a willful Stillness entertain,
> 'Screwing their Faces in a politick Form,
> 'To cheat Observers with a false Opinion
> Of Wisdom, Gravity, profound Conceit;
> As who should say, I am, Sir, an Oracle.

While much of Shakespeare's verse remains, much is changed, as the example shows (in deference to the reader's sensibility I refrain from quoting verse that is utterly Granville's[1]). Later, in the climactic trial scene, Granville introduces blatant melodrama. As Shylock prepares to exact his pound of flesh, Bassanio intercedes, first to offer himself to Shylock in place of Antonio; then, drawing his sword, to oppose the justice of the court. Thereupon Portia, disguised as the lawyer, appeals to the enraged Duke, who has ordered his arrest: 'Spare him, my Lord; I have a way to tame him. | Hear me one word' (p. 36); whereupon she presents her solution, prohibiting Shylock from shedding a single drop of Christian blood. And so the play proceeds to its conclusion in Act 5, emphasizing throughout (but well beyond Shakespeare's original) the themes of Love and Friendship.[2] Granville retained the ring episode but, not surprisingly, heavily rewrote it.

The Jew of Venice was the only version of *The Merchant of Venice* enacted in London for the next forty years. If we can judge by the number of performances, it was not, despite its concessions to prevailing taste, a very popular play.[3] Betterton

[1] In *Shakespeare from Betterton to Irving* (New York, 1920), 2 vols., i. 76–9, George C. D. Odell exercises no such restraint in his scathing analysis, and estimates that about one-third of the lines in the play are Granville's.

[2] See Spencer's introduction, *Five Restoration Adaptations*, 30–2, for a more detailed analysis of this and other aspects of Granville's adaptation.

[3] C. B. Hogan, *Shakespeare in the Theatre, 1701–1800* (Oxford, 1952), 2 vols., i. 461, records thirty-six performances through 1748; Ben Ross Schneider's index to *The London Stage, 1660–1800* (Carbondale, Ill., 1979), lists forty-three, but these include several that were not complete performances, such as one at the Blue Post Tavern, 18 Mar. 1724, part of a 'medley'

5. Macklin as Shylock, Mrs. Pope as Portia

played Bassanio, Thomas Dogget Shylock, and Anne Bracegirdle Portia (as we learn from the list of actors following the
Prologue written by Bevill Higgons, Granville's cousin). That
Dogget, a well-known comedian, was assigned Shylock's part
suggests that an earlier tradition—Shylock as a comic
villain—was being followed, or at any rate this was the
way the role was understood.[1] When Charles Macklin played
Shylock, using a text much closer to Shakespeare's, he too
was a villain, but a terrifying rather than a comic one, and
he was highly successful.[2] His production at Drury Lane

of theatricals. No performances are listed for 1706–11, and usually only one
or two a year after that, if any at all.

[1] See above, p. 10. The tradition resumed or begun by Dogget continued
when Benjamin Griffin, another comedian, took over the role, followed by
Anthony Boheme, John Ogden, Walter Aston, and John Arthur (Spencer, *Five
Restoration Adaptations*, 29). Compare A. C. Sprague, *Shakespeare and the Actors*
(Cambridge, Mass., 1944), 19, who claims that Dogget established a tradition
but did not follow one.

[2] Macklin possibly took his cue from Nicholas Rowe's remarks in the *Life*
of Shakespeare prefixed to his edition of 1709, cited in Brian Vickers (ed.),

numbered twenty-seven performances in the first year alone, with more than half of these following almost consecutively after the opening on 14 February 1741. With Macklin usually as Shylock, the play continued in the repertory almost uninterrupted to the end of the century as one of Shakespeare's most popular comedies.[1]

Macklin's 'restoration'

Macklin not only restored Morocco and Arragon, he also brought back Lancelot Gobbo and his father. As no copy of his prompt-book survives, we do not know precisely how much of Shakespeare's text was retained or omitted. Probably a good many lines were revised, and a year after the original production the composer Thomas Arne supplied Portia with a song, and Lorenzo sang two ditties that lasted through the century.[2] A good evening's entertainment, after all, was the object to ensure good box office receipts. Costumes followed contemporary fashion, in accordance with theatrical tradition

---

Shakespeare: The Critical Heritage, 5 vols. (1974–81), ii. 196: 'To these [i.e. Thersites in *Troilus* and Apemantus in *Timon*, 'masterpieces of ill Nature and satyrical Snarling'] I might add that incomparable Character of *Shylock* the *Jew*, in *The Merchant of Venice*; but tho' we have seen that Play Receiv'd and Acted as a Comedy, and the Part of the *Jew* perform'd by an Excellent Comedian [i.e. Dogget], yet I cannot but think it was design'd Tragically by the Author. There appears in it such a deadly Spirit of Revenge, such a savage Fierceness and Fellness, and such a bloody designation of Cruelty and Mischief, as cannot agree either with the Stile or Characters of Comedy.'

[1] Schneider's index lists no performances for only 1766 and at least one in 1758, 1764, 1765, 1785, 1793. Compare Hogan, *Shakespeare in the Theatre*, ii. 717, whose totals for each of Shakespeare's plays clearly indicate the *Merchant*'s popularity. At the rival theatre in Covent Garden, the play was also staged soon after Macklin's revival with various actors in the role of Shylock, e.g. James Rosco, Isaac Ridout, Lacy Ryan, none of whom enjoyed Macklin's success: ibid. i. 315–17.

[2] Odell, *Betterton to Irving*, i. 262. Lelyveld, *Shylock*, 21, notes that Nerissa and Jessica were also given songs. She concludes from Bell's edition (1774), which probably followed Macklin's script, that Macklin cut lines in many scenes (doubtless to make room for these and other additions), but the trial scene was kept almost intact and Act 5 'was entirely faithful to Shakespeare'. Like other actor-producers, Macklin continually tinkered with the text, which varied from production to production over the years. Odell laments, for example, that the Morocco and Arragon scenes were later cut, and Bassanio's casket scene 'curtailed beyond recognition, almost beyond the point of clarity' (*Betterton to Irving*, ii. 25–7).

since Shakespeare; not until the next century did historical authenticity overwhelm the staging of Shakespeare's plays. Macklin, however, reputedly attempted some authenticity: he wore a loose black gown, long wide trousers, and a red skullcap, which tended to emphasize the Pantalone rather than the traditional stage Jew, though for the part he also wore a wispy red beard.[1]

If Macklin's fierce, relentless Shylock dominated eighteenth-century conceptions of the role, it gave way in 1814 to Edmund Kean's equally original conception of Shakespeare's Jew as a man 'more sinned against than sinning';[2] in brief, a tragic figure. Kean's portrayal was startling in every respect. Having argued the management of Drury Lane into allowing him to essay the role, he stunned the rest of the cast with his representation. After a single rehearsal on the morning of the performance, 26 January 1814, he appeared on stage with a black wig (defying all stage tradition), loose gaberdine, and Venetian slippers. The boxes in the theatre were empty, and only about fifty people in the pit witnessed Kean's revolutionary interpretation; but theatre history was made that night.[3]

Nineteenth-century productions

Other actors since Macklin had played Shylock, including John Philip Kemble and George Frederick Cooke, but none had approached the role as Kean did. As against Kemble's 'artificiality' or Cooke's 'fiendish savagery', Kean 'intellectualized' his acting of Shylock, bringing a freshness and energy that was new.[4] It altered entirely William Hazlitt's view of Shylock, formed by earlier portrayals rather than Shakespeare's text.[5] On looking into what Shakespeare wrote and comparing it to

[1] Lelyveld, *Shylock*, 25. Cf. W. W. Appleton, *Charles Macklin: An Actor's Life* (Cambridge, Mass., 1960), 45-6.

[2] Influenced by Kean, Hazlitt borrowed Shakespeare's phrase from *King Lear* 3.2.60, to describe Shylock in *The Characters of Shakespeare's Plays* (1817), 269.

[3] F. W. Hawkins, *The Life of Edmund Kean*, 2 vols. (London, 1869; repr. New York, 1969), i. 124-37; for details of Kean's interpretation, see pp. 146-53.

[4] Lelyveld, *Shylock*, 43-9.

[5] See his reviews in the *Morning Chronicle* for 27 Jan. and 2 Feb. 1814, reprinted in *A View of the English Stage* (1818), 1-4.

Kean's representation, he found Shylock much more human than he had suspected or other actors had displayed. Commenting on this insight, he reveals moreover a bias not uncharacteristic of his time: 'Shakespeare could not easily divest his characters of their entire humanity; his Jew is more than half a Christian; and Mr. Kean's manner is much nearer the mark'.[1]

Although in the next decade William Charles Macready began performing in the role, he was never very happy with his portrayal of Shylock, however much both critics and public admired it. His productions, which eliminated many of the non-Shakespearian 'ditties' that had encumbered performances, emphasized Shakespeare's language. As against Kean's impassioned, intense portrayal of a persecuted martyr, Macready was dignified and stately, his Shylock a man consumed with malice.[2] Only Edwin Forrest's Shylock in America rivalled Kean's, though Samuel Phelps and Robert Campbell Maywood were also admired towards the end of that era.[3]

By now, elaborate stage settings and scenery had long since displaced Shakespeare's 'bare' stage. Restoration theatres had introduced movable, painted flats, and throughout the eighteenth century scenic effects became increasingly ingenious. In the next century, Macready tried in his productions to ensure that spectacle did not overwhelm performance. The stage design for *The Merchant of Venice* was, in fact, much admired by the *Times* reviewer as 'in the best possible taste, very beautiful, and yet nicely discriminated, so as not to overbalance the drama'. The moonlit garden in Act 5 was especially admired, 'sparkling with soft light, and melting away into a poetic indistinctness at the back'.[4] The comment is especially noteworthy as lately the play tended to end with Shylock's defeat in Act 4. By mid-century, however, the splendour, intricacy, and 'authenticity' of Charles Kean's productions became the dominant mode.

Since as an actor Charles Kean, unlike his father, was little more than competent, perhaps his staging of Shakespeare's

---

[1] Cited by Hawkins, *Life of Kean*, i. 137.

[2] Lelyveld, *Shylock*, 49–56.

[3] Ibid. 56, 67.

[4] Cited by Odell, *Betterton to Irving*, ii. 227.

plays helped compensate for other shortcomings. His productions were extremely elaborate, as the designs by Telbin reveal.[1] Using multiple levels and movable pieces, Kean transformed the stage of the Princess's Theatre into a Venetian carnival for the 1858 production of *The Merchant of Venice*. Music and dance concluded the elopement scene, which ended Act 2, while gondolas passed to and fro upon 'real' canals.[2] Claiming to adhere more strictly to Shakespeare's text, Kean restored Morocco and Arragon, who had again disappeared (in Kemble's and Phelps's scripts), but he had to shorten their speeches severely, as he did Bassanio's and Portia's in 3.2 and Lorenzo's and Jessica's in 5.1: scene-shifting and spectacular effects took time and effort, and some things had to give way. For similar reasons, he transposed or 'transfused' various scenes. Odell sums up Kean's version as 'a good acting play without undue favour to lines that happened to be poetical'. Or, as he puts it more generally, 'What Kean gave was nothing but Shakespeare; but alas! the great deal that he did not give was also Shakespeare. This however was expected in his day'.[3]

Kean and his wife, Ellen Tree, played Shylock and Portia, two of their best roles. Large numbers of dancers and 'supers' filled the stage, for Kean was especially adept at crowd management. Costumes reflected late sixteenth-century Italian fashions which, according to Kean's preface to his 1858 edition, chiefly derived from Caesar Vecellio's *Degli habiti antichi e moderni di diverse parti del mondo* (Venice, 1590). What in his father's performance had been a compelling drama became in the younger Kean's production 'a magnificent show'.[4] The set designer and scene mechanic rivalled the actor for pre-eminence—a phenomenon not unknown in today's theatre.

Henry Irving and Ellen Terry

This emphasis on the spectacular continued to the end of the century, which included the now legendary performances of

[1] See Nancy J. Doran Hazelton, *Historical Consciousness in Nineteenth-Century Shakespearean Staging* (Ann Arbor, Mich., 1987), 71–4.

[2] Peter Davison, Introduction to the Cornmarket facsimile of Kean's 1858 edition of *The Merchant of Venice* (1971), n.p.

[3] Odell, *Betterton to Irving*, ii. 296, 287.

[4] Lelyveld, *Shylock*, 59.

Henry Irving as Shylock. They began on 1 November 1879, at the Lyceum Theatre, with Ellen Terry as Portia, and ran for over 250 performances—an astonishing achievement for a Shakespeare revival.[1] As in Kean's productions, real palaces, real canals, real gondolas, and huge crowds decorated the stage, but the superlative acting of Irving and Terry nevertheless shone through.[2] To accommodate the spectacle but as much, I think, to highlight his acting, Irving heavily cut the text. Arragon once more departed, and other scenes were eliminated (e.g. 2.3., 3.5, 4.2) or curtailed (e.g. 1.2, 3.2, 5.1). But a new scene was added: Shylock returning home by lamplight in Act 2 and knocking on the door of his empty house—just once, in Irving's performance, which was highly moving, though later some actors allowed the moment to degenerate into a frenzy of despair.[3]

Irving's Shylock was 'the type of a persecuted race; almost the only gentleman in the play, and most ill-used'.[4] How far had interpretation proceeded since the days of Dogget or Macklin! Irving claimed to have taken his portrait from life, from observing a 'Moorish Jew' in Morocco.[5] He affected the audience accordingly:

His Jew was no doubt often repulsive, but he had moments of sheer humanity, when one felt with him and almost, or quite, suffered with him. Something of the eternal man, subject to the striving and suffering which is the common lot of all human beings, pierced through the crust of his greedy Jewishness, prey-demanding, revengeful, and bitter, and went to the heart. One almost forgave him.[6]

[1] Odell, *Betterton to Irving*, ii. 375.

[2] Ibid. 421-3; Lelyveld, *Shylock*, 81.

[3] For example, Herbert Beerbohm Tree (Lelyveld, *Shylock*, 100). For a detailed and well-documented analysis of Irving's production, see Alan Hughes, *Henry Irving, Shakespearean* (Cambridge, 1981), 227-41; and James Bulman, *Shakespeare in Performance: 'The Merchant of Venice'* (Manchester, 1991), 33-52. Reports of Irving's added scene vary, from a silent re-entry into the house to several knocks on the door: cf. e.g. Robert Hichins, 'Irving as Shylock', in H. A. Saintsbury and Cecil Palmer (eds.), *We Saw Him Act* (1939; repr. 1969), 168; Hughes, *Henry Irving*, 232; William Winter, *Shakespeare on the Stage* (New York, 1911), 186; Bulman, 38.

[4] Joseph Hatton, *Henry Irving's Impressions of America*, 2 vols. (London, 1884), i. 265; cited by Lelyveld, *Shylock*, 82.

[5] Hichins, 'Irving as Shylock', 167. Unlike other actors, who erroneously adopt a middle European accent, Irving continued a tradition of Shylock as an oriental Jew, as Antony Sher did in the 1987 RSC production (see below).

[6] Ibid.

6. Fitzgerald's drawing of Henry Irving as Shylock

Although other interpretations would later compete with Irving's, and from time to time he himself would alter some aspects, his basic conception remains even now a powerful influence in the theatre and in criticism.[1]

Ellen Terry's Portia was a match for Irving's Shylock, bringing to the role an emphasis on the heroine's womanliness: she was 'all grace, sparkle, piquancy, ardor, sweetness and passion'.[2] Before joining with Irving, she had played the role very effectively in the Bancrofts' production at the Prince of Wales Theatre in 1875 against a very poor Shylock (Charles Coghlan).[3] In 1868 she had acted with Irving in *Katherine and Petruchio*, but his engagement of her as Portia in *The Merchant of Venice* ten years later marked the beginning of a truly new and mutually beneficial relationship, which endured for over two decades. Choosing her for the leading

[1] For example, in Laurence Olivier's portrayal in 1970 (see below). Compare Bernard Grebanier, *Then Came Each Actor* (New York, 1975), 296.

[2] Ibid. 305.

[3] Odell, *Betterton to Irving*, ii. 306.

lady in his productions at the Lyceum was one of the wisest decisions Irving ever made.[1]

The role of Portia is fraught with pitfalls. She may be too spirited and flighty, as Kitty Clive was in Macklin's revival,[2] or too snobbish and sophisticated, as Joan Plowright was in Jonathan Miller's National Theatre production. Like other Shakespearian heroines who also adopt male attire for a while, she must be at once, or alternately, romantic, sensitive, witty, and intelligent—forceful but not overbearing, loving and in love but not sentimental.[3] Nerissa plays up to her in the early scenes, but the real test of her character lies in her relationship—what there is of it—with the convert, Jessica. Criticism of early productions is relatively sparse, unfortunately, but among the most successful Portias, besides Ellen Terry, were Peg Woffington and Sarah Siddons[4] in the eighteenth century, Helen Faucit, Helena Modjeska, and Julia Marlowe in the nineteenth.

Twentieth-century reaction

Before the twentieth century dawned, reaction had already set in against over-elaborate staging of Shakespeare's plays. Chief among the 'challengers' was William Poel, whose theories advocated a return to the simplicity and swiftness that characterized performances at Shakespeare's Globe, at least as he understood them. Furthermore, where *The Merchant of Venice* was concerned, he took great exception to the prevalent conception of Shylock as a tragic figure, a man wronged rather than a wrongdoer—a misconception Poel attributed, in part, to 'a change in a nation's religion or politics [which]

[1] See Austin Brereton, *The Life of Henry Irving* (1908; repr. New York, 1969), 262-7.

[2] Lelyveld, *Shylock*, 23. At the time, Clive was a celebrated comedienne; she may have been influenced, moreover, by the farcical element in Granville's adaptation.

[3] See Winter, *Shakespeare on the Stage*, 217-22, for Ellen Terry in the role, and reviews of her performances reprinted in Gāmini Salgādo, *Eyewitnesses of Shakespeare* (1975), 137-43.

[4] Her debut in London on 29 Dec. 1775 at Garrick's Drury Lane was as Portia, with Tom King as Shylock. Hampered by illness, she failed utterly, but redeemed herself in the role later on, in 1785-6. See Mrs Clement Parsons, *The Incomparable Siddons* (1909), 23-4, 125-6, and Roger Manville, *Sarah Siddons* (New York, 1971), 32-3, 131.

causes a change in the theatre'. Just as new plays are written to express new sentiment, old plays when revived are 'modified or readjusted' to bring them in line with new taste and opinion.[1] Arguing that Shylock is essentially a villain, despised not because he is a Jew but because he is a 'morose and malicious usurer', a 'curmudgeon' who, in a romantic comedy, must be defeated, Poel drew comparisons with Marlowe's Barabas and Molière's Harpagon to make his point. In Poel's view, Shylock's Jewishness is almost incidental; the play has more to do with his profession as usurer and rigid adherence to legalism than with religious convictions. Shakespeare's play was a protest against Marlowe's 'pagan Christians', not his 'inhuman Jew', Barabas.[2] The result of all this theorizing was a reversion to Shylock as a comic villain— complete with red wig—who at the end of the trial scene leaves the stage, not as a broken, pathetic figure, but as a man furious at having been outwitted.[3]

Other protests were sounded. An actress, who preferred to remain anonymous, published a critique of Portia, as played by Ellen Terry, that is really a polemic against the benevolent and impulsive warmth that had been so much admired, favouring instead a more tough-minded and resourceful character.[4] By the 1930s the Victorian tradition, which had reached its apex in Irving's production and was continued by Herbert Beerbohm Tree, Arthur Bourchier, Richard Mansfield, and others, had come to an end. Its last gasp was Frank Benson's final performance at the Shakespeare Memorial Theatre in Stratford-upon-Avon, 16 May 1932.[5] The new manager of the Festival, William Bridges-Adams, set out to breathe new life into Shakespearian productions. He invited

[1] William Poel, *Shakespeare in the Theatre* (1913), 70–1.

[2] Ibid. 71–84. Poel attributes some interesting political motives to Shakespeare here.

[3] See Lelyveld, *Shylock*, 97–8; Robert Speaight, *Shakespeare on the Stage* (1973), 136; and compare Poel, *Shakespeare in the Theatre*, 132, who, ignoring Shylock's last lines, justifies his conclusion by reference to the parallel episode in *Il Pecorone*.

[4] *The True Ophelia; and Other Studies of Shakespeare's Women*, by 'An Actress' (1913); quoted in Leigh Woods, *On Playing Shakespeare* (New York, 1991), 129–33.

[5] Bulman, 53, citing J. C. Trewin's account of the performance in *Shakespeare on the English Stage, 1900–1964* (1964), 137.

7. Komisarjevsky's set design (1932)

Russian-born Theodore Komisarjevsky to direct *The Merchant*, which opened on 25 July 1932. A 'cat-among-the pigeons' version, as J. C. Trewin says, it could hardly be more different from anything Stratford—or London—had previously witnessed.

Komisarjevsky was determined to overturn the tradition of 'pictorial realism' that, combined with historical detail, naturalistic acting, and moral sententiousness, had dominated productions for half a century or more.[1] While he kept all but fourteen lines of Shakespeare's text, restoring both princes, Morocco and Arragon, and both Gobbos, he added some mimed episodes, such as 'a capering of *commedia dell'arte* masquers' at the beginning, and a yawning Lancelot Gobbo at the end. Morocco, Arragon, and the Duke were burlesqued, and for the trial scene Portia wore bicycle-wheel spectacles and Antonio a large ruff that made his face look like the head of John the Baptist on a charger, according to one report.[2] For Komisarjevsky, *The Merchant of Venice* was a fantastic

[1] Bulman, 54.
[2] Trewin, *Shakespeare on the English Stage*, 137. Compare Bulman, 59–61, who notes that the leader of the opening masque, Bruno Barnabe, returned as Lancelot, his double role suggesting that the action of the play was Lancelot's dream.

comedy, to be staged as such. Sets, costumes, and acting expressed 'the emotional and rhythmic movement' of the play, not any particular period. Buildings veered off at odd angles, a Bridge of Sighs was split in two, and unearthly shades of green and crimson alternately bathed the stage to create 'a Venice of popular dreams'. Antonio, dressed in brilliant colours, became 'a depraved exquisite', the cause of his melancholy, self-love.[1] Into this extravaganza, Randle Ayrton's traditional Shylock was out of place, resisting Komisarjevsky's attempt to reduce the role to comic villainy.[2]

Despite the fantasy, or underlying it, Komisarjevsky's *Merchant* had a political and social goal. This was to satirize capitalist, bourgeois 'decadence' which, in his view, engendered racial prejudice. Komisarjevsky saw the young, effete Venetians as dissipated idlers and Shylock, however malicious, as someone suffering injustice at their hands.[3] To this extent, his conception was not altogether removed from nineteenth-century interpretations. Indeed, throughout its stage history, the play continues to oscillate between the poles of tragedy and comedy, justifying critical insistence on its ambivalent, or contradictory, nature. Following World War II, when the extreme of racial hatred stood revealed, directors found the Holocaust impossible to ignore, although their ways of dealing with the recent historical horror varied from production to production.

Jonathan Miller's way of dealing with it in his 1970 National Theatre production was to stage the play in an epoch closer to our own. His Venice, set in 1880, closely resembled mid-Victorian London. His Shylock, enacted by Laurence Olivier, resembled assimilationist Jews, such as Benjamin Disraeli (whose features Olivier emulated), bankers and financiers (not Elizabethan usurers), such as Baron Rothschild. Like Komisarjevsky, Miller had a social programme—to show the roots of modern anti-Semitism in economics and the competition for power, which have more to do with capitalism and politics than with biblical theories of the death of Christ.[4] To make

[1] Bulman, 56–8.

[2] Trewin, *Shakespeare on the English Stage*, 137; Bulman, 68–71.

[3] Bulman, 72.

[4] Bulman, 76–7, citing Miller's reference to Hannah Arendt, *The Origins of Totalitarianism*. Bulman's extended analysis of Miller's production is excellent.

8. Laurence Olivier as Shylock, Joan Plowright as Portia: National Theatre (1970)

Shylock a more sympathetic character, he heavily cut the text, excising for example Shylock's long aside, 1.3.38–49 ('I hate him for he is a Christian') and reducing or eliminating (in the televised version) Lancelot's low comedy. Further, he saw Jessica as a young woman deeply troubled by the desertion of her father and her religion. Shylock's reaction to her elopement—and to Salarino and Solanio's taunting—leads directly to his quest for revenge against Antonio as representative of the Christian community that has grievously wronged him. To emphasize the tragic aspect, so that an audience could not help grasping his intention, Miller ended the production with a startling *coup de théâtre*. As Portia and the others enter her house, Jessica drifts away in the opposite direction, holding the deed Shylock has signed, while offstage a cantor intones the mourner's *Kaddish*.

Ten years later, Miller again produced *The Merchant of Venice*, this time for the BBC television series of Shakespeare's plays. Constrained by series policy, this production was perforce more traditional. Jack Gold, who had little experience with Shakespeare, was the director. Warren Mitchell played Shylock; but whereas Olivier adopted the speech accents of a 'too consciously naturalised alien', appropriate for the interpretation he followed, Mitchell spoke with a 'thickish accent' of a middle European Jew, and both actors represented men more driven by emotion than concupiscence.[1] In the intervening period between the two productions, Miller had come to see the play as 'totally symmetrical in its prejudices'.[2] Both sides were in the wrong, as in the trial scene, which balances Shylock's inhuman behaviour towards Antonio with the brutality of his forced conversion.

The Royal Shakespeare Company's 'Merchants'

Meanwhile, another British director, John Barton, also staged two productions, an experimental one at The Other Place in Stratford-upon-Avon (1978), and a full-scale production in the main house (1981).[3] Both were set in the later nineteenth century, though less elaborately than Miller. Two quite different actors played Shylock for these productions, with two quite different conceptions of the role. Their differences further demonstrate the ambiguities inherent in the play. Patrick Stewart's Shylock at The Other Place was a mean-spirited man, interested more in money than anything else—love, family, religion, social status. For Stewart, Shylock was a bad man and a bad Jew; hence, he readily accedes to Antonio's demand at the end to convert to Christianity: no groans or moans accompanied his yielding, just a nervous, ingratiating giggle. This was totally in keeping with Barton's idea, that 'the play is about true and false value and not about race'.[4]

[1] Bill Overton, *Text and Performance: 'The Merchant of Venice'* (Atlantic Highlands, NJ, 1987), 49.

[2] Bulman, 101, quoting from the interview on PBS (the American Public Broadcasting System), 22 Feb. 1982. See also Marion D. Perret, 'Shakespeare and Anti-Semitism: Two Television Versions of *The Merchant of Venice*', *Mosaic*, 16 (1983), 145–63.

[3] Both afterwards transferred to London: the first to the Warehouse in 1979, the second to the Aldwych Theatre later the same year.

[4] Overton, 53, quoting from the 1981 theatre programme. See also Patrick Stewart, 'Shylock in *The Merchant of Venice*', in Philip Brockbank (ed.), *Players of Shakespeare* (Cambridge, 1985), 11–28, for the genesis and development of his portrayal.

9. Patrick Stewart as Shylock: Royal Shakespeare Company (The Other Place, 1978)

10. David Suchet as Shylock, Sinead Cusack as Portia, Tom Wilkinson as Antonio: Royal Shakespeare Company (1981)

For David Suchet, who like Warren Mitchell is himself Jewish, matters were more complicated. For him, Shylock's Jewishness is central; he is not an outsider who happens to be a Jew; he is an outsider *because* he is a Jew.[1] The difference was highlighted by Suchet's slight accent, occasioned by his sense of Shylock's pride of race, as against the absence of anything 'Jewish' in Stewart's speaking.[2]

Other differences emerged that are instructive. Suchet drew a distinction between the public Shylock and the private one. For him, the most difficult scene was 2.5, Shylock at home with Jessica. Suchet wanted to show Shylock's tender feelings towards his daughter and at the same time justify her desertion, and he could not reconcile the contradiction. Stewart showed no tenderness towards Jessica; on the contrary, at one point, sensing a flicker of defiance, he slapped her resoundingly across the cheek.[3] But for both actors, the decisive moment came at 3.1.118, when Shylock laments the loss of Leah's ring and determines then on his revenge.[4] As for Olivier's Shylock, Jessica's elopement was crucial, and 3.1 rather than the trial scene was the real climax of the drama.

Two more British productions may be briefly considered before going on to productions abroad. After John Caird's disastrous, overdesigned production in 1984, the Royal Shakespeare Company again staged the play in 1987, with Antony Sher as Shylock under Bill Alexander's direction. Sher reverted to an unassimilated, oriental Shylock, with a corresponding accent and costume, and Deborah Goodman's Jessica followed suit. The production nevertheless was not without contemporary allusions, such as swastikas and similar graffiti scrawled on walls near Shylock's home. In fact, the production deliberately intensified the problematic nature of the text, compelling the audience to examine the nature of their own prejudices. As a result, this *Merchant* became highly con-

[1] See John Barton, *Playing Shakespeare* (1984), 171. In this chapter, called 'Exploring a Character', Suchet and Stewart discuss their views of Shylock with Barton, and in the videotapes (from which the book was transcribed) they enact several key passages or scenes. Compare their interviews with Judith Cook, *Shakespeare's Players* (1983), 80–6.

[2] Overton, 53; compare Barton, *Playing Shakespeare*, 172.

[3] Stewart, 'Shylock', 22; Barton, *Playing Shakespeare*, 176.

[4] Barton, *Playing Shakespeare*, 177–8.

11. Antony Sher as Shylock, Deborah Goodman as Jessica: Royal Shakespeare Company (1987)

troversial, not least because, in stripping bare his bloodthirsty motives, Sher made Shylock highly offensive.[1]

Alexander did not want this production to be about anti-Semitism only but about racism generally, and as a South African Jew, Sher concurred.[2] Hence, in the trial scene, while commenting on Venetian slave-holding (4.1.89–99), Shylock held a black attendant before him, connecting discrimination against Jews and blacks and making it visually unmistakable.[3] On the other hand, Alexander felt strongly that the play should be set in a Jacobean period (1630) to emphasize the cruelty and rigorous justice of that world. Only thus, he believed, could Shylock's intention to carve the heart out of Antonio be credible; to put it in a Victorian or later context would lessen credibility. Moreover, the social context requires the historical setting, he argued, to understand the position of Jews in Venice and Christian hypocrisy in dealing with them.[4]

[1] Bulman, 117–20.

[2] See Antony Sher, 'Shaping up to Shakespeare', *Drama*, 4 (1987), 28; cited by Bulman, 120.

[3] Bulman, 124–5.

[4] See the interview with him in Ralph Berry, *On Directing Shakespeare* (1989), 181–2.

Other aspects of Alexander's production were also calculated to dislodge audience complacency and, indeed, arouse controversy, such as the blatant homosexuality Antonio revealed in his relationship with Bassanio. But Sir Peter Hall's production at the Phoenix Theatre in 1989, which borrowed a good deal from Alexander's, seems to have occasioned comparatively little uproar, even in New York, where it was staged the following year. Perhaps critics and public had grown weary of debate, or the presence of another, more popular Jewish actor, the American Dustin Hoffman, defused controversy while it promoted box office success. More to the point, Hall's production, compared to Alexander's, was milder in tone, certainly less tendentious, though not less elaborately staged, and a great deal of spitting by Christians occurred in both. In his first Shakespearian role, Hoffman was hardly a dominating presence, giving Geraldine James's Portia an opportunity to shine—as she did—thereby effecting a necessary and useful counterbalance to Shylock. And like Sinead Cusack in Barton's 1981 staging, she seemed to be the only one with any compassion for Shylock at the moment of his forced conversion.[1]

The Merchant of Venice *abroad*

Dustin Hoffman's Shylock is a reminder that *The Merchant of Venice*, like all of Shakespeare's plays, does not belong exclusively to the English theatre; as part of the world's, it has been performed in all corners of the earth. It was the first of Shakespeare's plays to be performed by professional actors in America, when Lewis Hallam and a company from London staged it in Williamsburg, Virginia, on 15 September 1752.[2] By far the greatest American actor to play Shylock was Edwin Booth, who succeeded his father, Junius Brutus Booth, in the

[1] On the London production, see Stanley Wells, 'Shakespeare Production in England in 1989', *SSur* 43 (1991), 187–8; for the New York production, see Irene Dash, '*The Merchant of Venice*', *Shakespeare Bulletin*, 8 (Spring 1990), 10–11, and Bernice W. Kliman, 'The Hall/Hoffman Merchant: Which is the Anti-Semite Here?', *Shakespeare Bulletin*, 8 (Spring 1990), 11–13.

[2] Charles Shattuck, *Shakespeare on the American Stage: From the Hallams to Edwin Booth* (Washington, DC, 1976), 3. But cf. Hugh F. Rankin, *The Theatre in Colonial America* (Chapel Hill, NC, 1960), 191, who notes performances of *Richard III* and *Othello* that antedate *Merchant* (cited by Shattuck, 15).

role. By the early nineteenth century, the stock figure of the Jew was well recognized, and the elder Booth, who knew Hebrew, played Shylock using a Yiddish accent.[1]

Edwin Booth made his debut in London as Shylock in 1861 and later not only surpassed his father but another eminent American actor in the role, Edwin Forrest. Whereas Forrest's representation rivalled Kean's, the younger Booth's more closely resembled Macklin's or, in our own time, Patrick Stewart's. For he saw Shylock as essentially driven by economic concerns—an avaricious old man devoid of love or compassion, moved more by the 'money value' of Leah's ring than by sentimental associations.[2] For his revivals of the play in New York beginning in 1867, which none surpassed until Henry Irving's, he cut or otherwise altered the text, making Shylock's threats more ominous and bringing the role into still greater prominence. He ended 1.3, for example, after Antonio and Bassanio go off, with: 'Thou called'st me dog before thou hadst a cause, | But since I am a dog, beware my fangs' (3.3.6–7), and concluded the play with Shylock's exit in 4.1. He costumed Shylock as an oriental patriarch whose appearance was 'grotesque' but also tragic, and acted him accordingly.[3]

In the nineteenth century another, less well-known American actor played Shylock with great success. Since he was black, his appearances occurred mainly in Europe, where he won great acclaim, especially in Russia—not as an oddity, but for his stirring portrayal of many Shakespearian characters, King Lear as well as Shylock. This actor was Ira Aldridge, born in New York, 24 July 1807, and dead at 60 in Lodz, Poland. His Shylock reminded the critic K. Zvantsev more of the Wandering Jew than Shakespeare's money-lender, but the comment reveals Aldridge's attempt to universalize the character, making him representative of diaspora Jews generally.[4] More recently, at Stratford, Connecticut, in 1957, Morris Carnovsky starred as Shylock. Acclaimed by critics as a

[1] Lelyveld, *Shylock*, 63–5.

[2] See Booth's analysis of the role in Furness, 383–4, and compare Winter, *Shakespeare on the Stage*, 153–9.

[3] Winter, *Shakespeare on the Stage*, 155–9.

[4] See Herbert Marshall and Mildred Stock, *Ira Aldridge : The Negro Tragedian* (1958), 235.

'superb' Shylock, particularly for his vigour and humour,[1] he recognized the 'intellectual' quality of Shylock's language and its 'magnificent, haunting diction', which he refused to debase with an accent, as the fashion was then in England.[2] At about the same time, at the Oregon Shakespeare Festival, *The Merchant of Venice* was staged in 'authentic' Elizabethan fashion, with Angus Bowmer, the actor and founding director, playing Shylock as a comic villain, complete with red wig and beard, putty nose, and middle European accent. For someone (like this editor) brought up on 'tragic' Shylocks, Bowmer's rendition was a shock—and a revelation. For it worked; it helped realize the play's comic structure and intent, without at the same time provoking repugnant anti-Semitic responses. A quite different, modern dress version was performed on the same stage in 1991, a controversial production that did, unfortunately, arouse spirited accusations of anti-Semitism.[3]

Elsewhere, *The Merchant of Venice* has been played in a variety of styles with a variety of emphases. Not surprisingly, during the Nazi era in Germany, it was used along with Marlowe's *Jew of Malta* mainly for propaganda purposes to ridicule Jews, who were then being systematically eliminated for a *Judenfrei* Third Reich. It was not always thus, as in Max Reinhardt's splendid productions in Germany and in Italy (1905-35), with Rudolf Schildkraut, Albert Basserman, and Memo Benassi playing Shylock.[4] And it was not in the 1963 Berlin production, directed by Erwin Piscator, with Ernst Deutsch as a very human Shylock; or in the 1968-9 Austro-German television co-production, with Fritz Kortner as a less dignified but more realistic Shylock.[5]

Israeli Productions

Although *The Merchant of Venice* has not been performed as frequently in Israel as in other countries, its several produc-

[1] Lelyveld, *Shylock*, 114.

[2] Morris Carnovsky, 'On Playing the Role of Shylock', in Francis Fergusson (ed.), *The Merchant of Venice* (New York, 1958), 25.

[3] See e.g. Glenn Loney, 'P.C. or Not P.C.', *Theatre Week*, 2-8 Sept. 1991, pp. 29-31.

[4] Speaight, *Shakespeare on the Stage*, 206-8; J. L. Styan, *Max Reinhardt* (Cambridge, 1982), 61-4.

[5] See Bulman, 151-3, for an account of George Tabori's production in

tions there have aroused considerable interest. The first, in 1936, before the State of Israel was formed, was at the Habimah Theatre, directed by Leopold Jessner, a Jewish refugee, famous for his work at Berlin's Staatstheater and the Schiller Theater in the 1920s. His Shylock, played alternately by Aharon Meskin and Shim'on Finkel, stood for all the Jewish people battling with Christian society. Meskin emphasized Shylock's heroic stature, Finkel his bitter spite, and every trace of comic villainy was banished. Nevertheless, because the Christians appeared too decent, the production evoked controversy. A public 'trial' was organized by and held at the theatre, where the author, the director, and the theatre were all defendants charged with fomenting anti-Semitism.[1]

Tyrone Guthrie directed the next production in Israel many years later, in 1959, well after the Nazi Holocaust had ended and the State of Israel was established. Meskin again played Shylock in this modern dress production, where he affected a Rothschildian appearance to underscore Shylock's position as a financier. Guthrie attempted to keep the play within the bounds of romantic comedy, but his attempt to present a 'fantasia on the twin themes of mercy and justice' was only partly successful.[2] His Portia was miscast, and Meskin's marked pathos fitted awkwardly into the romantic conception. After a few months, the production disappeared from the Habimah's repertoire, and the play was not staged again until 1972, when Israeli-born Yossi Yzraeli directed it at the Cameri Theatre in Tel Aviv. By this time, after the stunning victory of the Six-Day War in 1967, Israeli pride and self-confidence were such that an unsympathetic Shylock could be risked in a version far removed from realism. A mimed Good Friday procession opened the play; in the trial scene Antonio appeared as a Christ figure, with a large black cross fitted on his back; and throughout the performance a puppet theatre

Munich, 1978, and elsewhere, and Maria Verch, 'The Merchant of Venice on the German Stage since 1945', Theatre History Studies, 5 (1985), 84–94.

[1] See Avraham Oz, 'Transformations of Authenticity: The Merchant of Venice in Israel 1936–1980', in Werner Habicht (ed.), Shakespeare Jahrbuch West (Bochum, 1983), 165–77. Subsequent references to Israeli productions during this period are to this article.

[2] Tyrone Guthrie, In Various Directions: A View of the Theatre (1965), 103; cited by Oz, 'Transformations', 173.

in the background mimicked the action of the main stage or otherwise commented upon it. A noted Israeli comedian, Avner Hyskiahu, grotesquely impersonated Shylock as a shrewd old Jew used to making clever deals. Not surprisingly, the production was a box office and artistic failure that aroused much hostile criticism.

In 1980 the Cameri Theatre again attempted the *Merchant*, importing Barry Kyle from the RSC to direct, with Christopher Morley as set designer. To placate or rather defuse anticipated objections, some of the cast persuaded Kyle to delete Shylock's conversion in 4.1. Kyle's conception of Shylock as a man 'succumbing to the logic and mentality of terrorism', however, proved not much more successful than other versions of this controversial character, especially in a country which by then had long occupied another people's territory. In this political and social context, Kyle's message of 'concord and love' was not as fully appreciated as he and others hoped. In 1986 the Bet-Lessin Theatre in Tel Aviv attempted a revival in a new translation by Avraham Oz, directed by Yossi Alfi, but the production was stopped during rehearsals for want of sufficient financing. Two years later, Moshe Shamir adapted the play as *A Carnival in Venice*, produced at the Festival of Alternative Theatre in Acre.

The Merchant of Venice continues to hold the stage the world over. The reasons for its popularity remain the same: its rich complexity, which both actors and critics continue to explore rewardingly, and its marvellous opportunities for acting, staging, and set design. It still arouses controversy, especially in audiences sensitive to the elements of anti-Semitism it undoubtedly contains. Whether the play is itself anti-Semitic or not depends largely upon one's interpretation, on the stage as on the page, and one's inclination or not to accept the ambiguities, inconsistencies, and contradictions inherent in the text. Resolving those ambiguities may predispose representation in one direction or another, as we have seen, but it may also diminish the play that Shakespeare wrote.

The most excellent

Historie of the *Merchant*
of *Venice*.

VVith the extreame crueltie of *Shylocke* the Iewe
towards the sayd Merchant, in cutting a iust pound
of his flesh: and the obtayning of *Portia*
by the choyse of three
chests.

As it hath beene diuers times acted by the Lord
Chamberlaine his Seruants.

Written by William Shakespeare.

AT LONDON,
Printed by *I. R.* for Thomas Heyes,
and are to be sold in Paules Church-yard, at the
signe of the Greene Dragon.
1600.

12. Title-page of the 1600 Quarto

TEXTUAL INTRODUCTION

On 22 July 1598, the following entry was recorded in the Stationers' Register:

Iames Robertes. | Entred for his copie vnder the handes of bothe the wardens, a booke of the Marchaunt of Venyce or otherwise called the Iewe of Venyce. | Provided that yt bee not prynted by the said Iames Robert*es* or anye other whatsouer w^{th}out lycence first had from the Right honorable the lord Chamberlen   vj^d

The explicit provision, that the book was not to be printed by Roberts or any other printer without the Lord Chamberlain's licence, implies that this was essentially a 'blocking', or 'staying', entry used to establish copyright but at the same time to prevent publication until such time as the players, or the Lord Chamberlain, their patron, gave permission.[1]

Two years later, on 28 October 1600, another entry appeared in the Register:

Tho. Haies Entred for his copie vnder the hand*es* of the Wardens & by Consent of m^r Robert*es* A Booke called the booke of the m chant of Venyce   vj^d

The play was later printed in the same year by Roberts for Hayes with the following title-page:

The most excellent | Historie of the *Merchant* | *of Venice*. | VVith the extreame crueltie of *Shylocke* the Iewe | towards the sayd Merchant, in cutting a iust pound | of his flesh: and the obtayning of *Portia* | by the choyse of three | chests. | *As it hath beene diuers times acted by the Lord* | *Chamberlaine his Seruants.* | Written by William Shakespeare. | [Printer's ornament] | AT LONDON, | Printed by *I.R.* for Thomas Heyes, | and are to be sold in Paules Church-yard, at the | signe of the Greene Dragon. | 1600.

Although further details have not been forthcoming regarding the transactions among Roberts, Hayes, and the Lord Chamberlain's Men, the second entry in the Stationers' Register probably signifies that Roberts now demonstrated before the

---

[1] W. W. Greg, *The Editorial Problem in Shakespeare* (Oxford, 1942), 123.

wardens that he had permission to publish the play. At the same time, Roberts (acting for himself or on behalf of the Lord Chamberlain or his servants, the players) transferred his right to Thomas Hayes, who soon published the book that Roberts printed for him. The curious redundancy, 'A booke called the booke . . .', suggests it was the prompt-book that was produced before the wardens, since prompt-books were usually inscribed as 'The book of . . .' and the scribe apparently copied the title he saw before him on the manuscript.[1]

The copy actually used in printing the first quarto was probably not the prompt-book but some manuscript, very likely a fair copy, possibly in Shakespeare's own hand.[2] The presence of a number of indeterminate or descriptive stage directions points away from the prompt-book, while the absence of numerous printing errors, which may derive from heavily revised foul papers, points away from foul-paper provenance.[3] For example, at 2.1.0 the stage direction reads: 'Enter *Morochus* a tawnie Moore all in white, and three or foure followers accordingly, with *Portia*, *Nerissa*, and their traine', and at 3.2.62: '*A Song whilst Bassanio comments on the caskets to himselfe*'. Again, the stage directions at 2.2.106, 2.7.0, 3.2.0, 4.1.0, and elsewhere indicate an indeterminate number of followers. In several places entrances or, more often, exits are missing, e.g. at 2.3.14, 2.6.50, 4.2.0. Some may be clearly suggested by the dialogue, but entries at least would have been needed in a prompt-book (though not necessarily supplied). The fact that additional music cues, presumably deriving from a prompt-book, are given in the Folio text also suggests that Q does not derive from a prompt-book.

On the other hand, a few imperative stage directions read like those found in prompt-books, e.g. '*open the letter*' (3.2.234), '*play Musique*' (5.1.68); but here the author, who was after all a professional playwright, might simply be anticipating the bookkeeper.[4] Similarly, explanatory directions,

---

[1] See Greg, *SFF* 256.

[2] Mahood feels fairly certain that it was. See NCS 172–3.

[3] Brown, p. xiv; compare NCS 170, where the argument for scribal fair copy is reviewed. Cf. William B. Long, 'Stage Directions: A Misrepresented Factor in Determining Textual Provenance,' *TEXT* 2 (1985), 121–37.

[4] See Greg, *Editorial Problem*, 36; Brown, p. xv.

e.g. '*Enter Iewe and his man that was the Clowne.*' (2.5.0), and '. . . *Salerio* a messenger from Venice.' (3.2.217), do not suggest a bookkeeper's annotations.[1] Lancelot's entrance and dialogue, 5.1.39–48, which Greg once suggested could have been an interpolation inserted into the prompt-book, may just as well have been inserted in foul papers, as Greg later realized; or it may not have been an interpolation at all.[2] Far more intriguing is the mystery of the three 'Sallies', John Dover Wilson's nickname for Salarino, Solanio, and Salerio. Most recent editors follow Wilson and reduce these to two, Solanio and Salerio, since Salarino might simply be a diminutive or corruption of Salerio.[3] The evidence is often confusing and ambiguous, and the arguments against three Sallies are far from conclusive, as M. M. Mahood has shown.[4] Particularly striking is the specific identification of Salerio as 'a messenger from Venice' (3.2.217), which would seem superfluous if he was one of the Sallies introduced earlier, who had appeared in the immediately preceding scene.[5] Since doubling was commonplace (indeed, the rule), the actor who played Arragon or Old Gobbo could as easily play Salerio, someone distinguishable (if just barely) from the other two Sallies.[6] This edition, therefore, follows Mahood in retaining Salarino.[7]

[1] Compare Greg, *SFF* 257.

[2] See Commentary at 5.1.49.

[3] NS 100–4; Greg, *SSF* 257–8, Brown, 2 n. Cf. Sisson, 135, who believes that the three Sallies were reduced to two, as Wilson believed, in a playhouse revision.

[4] NCS 179–83. Note especially the chart listing all the speech headings in Q, Q2, and F on pp. 180–1.

[5] See Collation for 3.3.0. Q2 alters Q's 'Salerio' to 'Salarino', F to 'Solanio'. As Professor George Walton Williams says (in a private communication), 'There is nothing in the fiction that suggests that Salerio of 3.2 should have returned to Venice to appear in 3.3, and there is everything in theatrical practise that prohibits it.' Not overly concerned about minor characters, Shakespeare probably confused their names. Since Salerio is repeatedly so called in the dialogue (e.g. 3.2.218, 226, 236, 264), he is the one with the greatest certainty. On the other hand, again as Professor Williams suggests, there could be two characters named Salerio, just as Balthasar is used twice in this play and Jaques twice in *As You Like It*. (Cf. NCS 179.)

[6] NCS 179.

[7] Note, too, that although abbreviated speech prefixes are notoriously unreliable, 'Salarino' is several times distinctly spelled out in stage directions, e.g. at 1.1.0, 1.1.68.1, 2.4.0, 2.6.0, 2.8.0, 3.1.0.

The printing of Q

The first quarto was printed in James Roberts's printing house by the same two compositors who set the second quarto of *Titus Andronicus* (1600) and later set the second quarto of *Hamlet* (1604/5), as Brown has demonstrated from the evidence of spelling tests, the length of running titles, use of italic type, indented lines, etc.[1] Both were competent workmen who carefully followed copy. The differences between Q2 *Hamlet* and Q *Merchant* derive not so much from the compositors' ability or willingness to follow copy as in the nature of the copy itself. For Q2 *Hamlet* it was heavily revised holograph that the compositors had difficulty reading; consequently, many errors resulted.

According to Brown's tabulations for Q *Merchant*, Compositors X and Y (as he called them) set the pages on the following sheets:

Compositor X: A1 (title-page), C, E, G, I, K
Compositor Y: A2–4^v, B, D, F, H

In other words, Compositor X set 2.1.17–2.5.3, 2.9.27–3.2.102, 3.4.19–4.1.141, 4.1.426–5.1.307, and Compositor Y set all the rest.[2] In stints by each compositor the absence of capital letters, mainly at the beginning of verse lines, becomes apparent sporadically in sheets A–C, more frequently and consistently thereafter. A shortage of capital letters in the type case probably best explains this peculiarity in Q.[3] Another peculiarity is the abundance of question marks, mainly on sheets G, I, and K, all set by Compositor X, but not on sheets C and E, also set by him. Again, a shortage of type may explain this, but since question marks sometimes substituted for exclamation points, we cannot always be sure whether a period or an exclamation point was intended where a question

---

[1] 'The Compositors of *Hamlet* Q2 and *The Merchant of Venice*', SB 7 (1955), 17–40.

[2] Brown, 'Compositors', 31; TC 323.

[3] TC 324; Cf. NCS 172. Mahood suggests that Shakespeare may have stopped capitalizing the first letters of some verse lines, as in the holograph portion (Hand D) attributed to him in *The Book of Sir Thomas More*, and the compositors followed copy. It seems more likely that Shakespeare did not capitalize first letters of verse lines at all, and the compositors, for whatever reason (forgetfulness, shortage of type), followed copy instead of convention.

mark is clearly wrong.[1] Where an exclamation point or other mark of punctuation substitutes in this edition for a doubtful question mark in Q, the collation records the change, particularly from 3.5 to 5.1, where a large number of alterations are necessary.

The shortage of certain capital letters may have led to other peculiarities as well, such as the otherwise meaningless alternation of some speech prefixes, specifically *'Iewe.'/'Shy(l).'* and *'Clowne'/'Lau(nce)'*.[2] In as yet unpublished papers, Richard F. Kennedy has studied these alternations in minute detail as well as comparable ones in *A Midsummer Night's Dream* and has concluded that it was indeed a shortage of capital *I*s, in both roman and italic type, that led the compositors to switch speech prefixes from *'Iew(e).'* to *'Shy(l).'*.[3] Underlying his analysis is the assumption that the speech prefixes in copy were normally and consistently *'Iew(e).'*, an assumption which of course cannot be verified. Possibly they were, although the author as well as the compositors may have occasionally been led to adopt *'Shy[l].'* as a speech prefix after the name appeared in the dialogue.[4] One example (among others) occurs at 1.3.49–50, where the speech prefix *'Shyl.'* immediately follows Bassanio's *'Shyloch, doe you heare.'*, and a swash 'I' is the first word after it. On the other hand, the compositors for Q2 and F did not follow copy in this regard, either. Although the press reader or editor for Q2 seems to have tried to designate all speech prefixes consistently as 'Shy.', consistency breaks down from 3.3.1 onwards.[5] As shortage of capital letters in Jaggard's printing shop is unlikely, it may be that

[1] A shortage of colons as well as periods may also have existed. Compare *TC* 324; NCS 172.

[2] NS 94–5; cf. Brown, p. xvii n. 4.

[3] Professor George Walton Williams kindly sent me the paper mostly on *Dream*, which was delivered at his 1990 SAA seminar, and Professor Kennedy kindly sent me a draft of his paper on *Merchant*.

[4] Compare Spencer, *Genesis*, 92–5, who believes the alternation of speech prefixes derives from the way Shakespeare regarded Shylock at any given moment.

[5] See the chart recording these speech prefixes in the Appendix. Similar alternations occur among the seven stage directions for Shylock. The first is *Enter Shylocke the Iew.* (1.3.0), then *Enter Iewe ...* (2.5.0), *Enter Shylocke* (3.1.20.1), *Enter the Iew ...* (3.3.0), *Exit Iew.* (3.3.17), *Enter Shylocke* (4.1.14), and finally *Exit.* (4.1.396). Q2 and F duplicate these directions almost exactly in every instance.

'Jew' and 'Shylock' became synonymous for those engaged in printing the text as well as for the author. Again, although Lancelot Gobbo is always called Lancelot in the dialogue, in the stage directions and speech prefixes he is referred to as '*Clowne*', except in sheet C, where the outer forme has '*Launce.*' (or some derivative) thirteen times and the inner forme has nine. Kennedy explains this variation by assuming that Compositor X, who set C, had a shortage of C's and an abundance of L's. The shift in Lancelot's surname from *Iobbe* (2.2.3–8) to *Gobbo* (2.2.29 SD–161 SD), however, is otherwise explained as possibly reflecting a change of intention on Shakespeare's part. He may have begun by using the Italian form of 'Job', then switched to the funnier sounding 'Gobbo', with or without the intention of giving the Gobbos a hump (in Italian, *gobbo* means hunchback: see Commentary on 2.1.1).[1]

The Pavier Quarto (1619)

The first quarto of *The Merchant of Venice* was reprinted in 1619 as part of an intended collection of Shakespeare's plays printed in William Jaggard's shop for Thomas Pavier.[2] When the collection was aborted, a false title-page with the date 1600 (but without Hayes's name) was printed so that the quarto might pass as the first edition, just as the Pavier quarto of *King Lear* used a false title-page and date for the same purpose. Although the Pavier quarto (Q2) has no clear textual authority, it shows many signs of having been more or less carefully edited. Some obvious errors are corrected, but other alterations in the text reveal a somewhat officious editor, revising or rewriting words or phrases. For example, in 1.3 he made the following changes, among others: 'although' for 'albeit' (58), 'are you resolu'd, | How much he would haue?' for 'is hee yet possest | How much ye would?' (61–2), and 'In th'end of Autume' for 'In end of Autume' (78). At 2.2.20 he corrects Q and has Lancelot's conscience and the fiend

---

[1] Compare Sisson, 135, who notes that 'I' and 'G' were frequently interchangeable in spelling. He cites as examples *Garret : Iarret, Genevora : Ienevora,* and *Giacomo : Iachimo,* but most of these involve a soft initial consonant before a vowel other than 'o'.

[2] See Greg, *Editorial Problem,* 131–4.

both counselling him 'well', but a few lines later he ruins a malapropism, changing 'deuill incarnation' to 'diuell incarnall' (25). He adds stage directions where some are missing, as at 2.2.113, and alters others, as at 2.5.0, where the explanatory '*Enter Iewe and his man that was the Clowne.*' becomes '*Enter the Iew and Lancelet.*'

Occasionally, the Q2 editor's corrections are important and anticipate those in the Folio, e.g. 'Slubber' for 'Slumber' (2.8.39), and at 3.3.0 he apparently recognized the need to alter '*Salerio*' to '*Salarino*'. He corrects the verse at 2.5.52–4, but incorrectly lines Old Gobbo's prose as verse at 2.2.47–9. Although on balance it seems unlikely that the editor collated his copy of Q against an authoritative manuscript of some sort, these and other substantive variants are recorded in the collation to this edition, since in at least a few instances, as in the changes '*Iobbe*' to '*Gobbo*' (2.2.3–8), '*Salerio*' to '*Salarino*' (3.3.0), and the speech prefixes '*Iewe.*' to '*Shy.*', (4.1.64, 66, 67), he may have had recourse to something other than his own judgement or guesswork.[1]

Many Q2 variants, however, may be the responsibility of Jaggard's compositor, possibly Compositor B, one of the two who worked on the Folio text of this play and who is notorious for his somewhat cavalier attitude towards setting plays in type.[2] If this is true, then to the compositor a number of errors or alterations may be attributed, such as dropped or added words, changes in word order,[3] and the omission of a speech prefix and line at 2.6.67. But the compositor may also have been responsible for some corrections, such as capitalizing the beginnings of verse lines, improved punctuation, modernized spellings, and the like. In many minor details,

---

[1] Compare Mahood, NCS 175–7, who gives a fuller listing of such changes and, while not attributing them to an authoritative source, regards them as important historically for showing 'how a Shakespeare play was read in 1619'.

[2] See D. F. McKenzie, 'Compositor B's Role in *The Merchant of Venice* Q2 (1619)', *SB* 12 (1959), 75–90, and Charlton Hinman, *The Printing and Proof-Reading of the First Folio of Shakespeare*, 2 vols. (Oxford, 1963), i. 10–12. Compare Peter Blayney, 'Compositor B and the Pavier Quartos: Problems of Identification and Their Implications', *Library*, 5th series, 27 (1972), 179–206, esp. 203–5. Blayney believes that two compositors, G and H, set Q2 *Merchant*, and that G's work somewhat resembles B's.

[3] NCS 174–5 gives a representative sampling of these.

however, it is not always possible to distinguish among the editor's alterations and the compositor's or even the author's.

The Folio 'Merchant'

The Merchant of Venice was again printed in the Folio of 1623 (F) from a copy of Q having the inner forme of sheet G in the uncorrected state (Q2 was printed from a copy of Q with the forme in the corrected state). That Q was used as copy for F is clear from the number of errors and other peculiarities common to both texts.[1] It was edited, however, to some extent by reference to a playhouse prompt-book, as the evidence of additional stage directions specifying flourishes, music, properties, etc., shows. Collation with the manuscript was undoubtedly cursory and perfunctory since a number of confusions persist that a prompt-book would normally have clarified, such as the problem involving the three 'Sallies'.[2] One of the music cues, moreover, is misplaced (see Commentary on 2.7.0). On the other hand, in view of James I's accession to the throne, Portia's denigrating reference to 'the Scottish lord' was carefully altered to 'the other Lord' (1.2.74), and other changes were made to conform to new laws regarding profanity (e.g. at 1.2.107, 5.1.157). Interestingly, though the two texts were printed in the same shop, the variants introduced into Q2 (1619) do not reappear in F (1623).[3]

As in Q2, the Folio editor and/or compositor corrected some errors while introducing others and in general made the punctuation somewhat heavier, modernized the spelling, and normalized grammar. Substantive variants involving changes in wording, such as 'smal' for 'meane' (1.2.7) and 'endlesse' for 'curelesse' (4.1.141), may have been introduced at some point in the prompt-book, but when and by whom is not known. Having doubtful authority, they have not been adopted in this edition where Q's meaning is sufficiently clear but are recorded in the collation.[4] Where F corrects Q exactly

[1] NCS 177 reprints the relevant passages.

[2] Mahood notes that the Q2 editor more carefully distinguished between the speech headings for Salarino and Solanio in 1.1 than the editor or (as she argues) the compositor of F did (NCS 178). Cf. Brown, p. xx.

[3] See Greg, *SFF* 261 n. B; *TC* 323.

[4] Possibly such changes were introduced in the prompt-book as concessions to easier audience comprehension, as some changes from Q to F *Lear* reveal

as Q2 does, e.g. changing 'Slumber' to 'Slubber' (2.8.39), it may be no more than a remarkable coincidence. But since Q2 was printed in Jaggard's shop, a copy of it could have been available for consultation when the editor or compositor needed to clarify something that seemed obscure in Q. Such consultation—if it occurred—was at most occasional and certainly not systematic.[1]

Besides numbering scenes, F added act divisions that most modern editions follow. Three compositors, B, C, and D, set the text,[2] which appears between *A Midsummer Night's Dream* and *As You Like It* on pages 163 to 184.[3] Pages 164 (O4ᵛ) and 165 (O5ʳ) are erroneously numbered 162 and 163; similarly, the penultimate page of *A Midsummer Night's Dream* (O3ʳ) is misnumbered 163, all resulting from an irregularity in the order of printing formes for quire O.[4] The Folio was reprinted in 1632, 1663, and 1684, but none of these editions has any textual authority, deriving as they do from reprints of what is essentially a reprint of Q. In 1637, Q was reprinted a third time, by Thomas Hayes's son, Laurence. His edition (Q3) is actually a more faithful reprint of Q than either Q2 or F, despite a number of typographical and other errors. Where Q2 includes 101 substantive variants from Q and F 92, Q3 has only 40.[5] But Q3 has no textual authority and therefore is not used in the collations in this edition.

(e.g. 'stick' for 'rash' and 'sterne' for dearne', 3.7.56, 61). But whereas F *Lear* shows many signs of authorial revision, F *Merchant* does not.

[1] Compare Brown, pp. xix-xx.

[2] For the assignment of compositors to pages set, see *TC* 149, and compare Hinman, *Printing*, II. 422–38.

[3] These consist of part of quire O (O4ʳ–O6ᵛ), all of quire P, and part of quire Q (Q1ʳ–Q2ᵛ).

[4] Hinman, *Printing*, II. 423–4.

[5] As calculated by Christopher Spencer, 'Shakespeare's *Merchant of Venice* in Sixty-Three Editions', *SB* 25 (1972), 104. Elsewhere in his article, 89–106, Spencer charts the history of editorial intervention over the centuries, noting the heavy dependence of modern editions on those that appeared before 1800. Editors who introduced the most new readings were Rowe, Capell, Pope, and Theobald, in that order. His data further confirm Fredson Bowers's judgement that the trend among recent editions is increasingly towards the restoration of original readings and less in the direction of independent emendation ('Today's Shakespeare Texts, and Tomorrow's', *SB* 19 (1966), 43; cited by Spencer, *Sixty-Three Editions*, 103 n. 18).

EDITORIAL PROCEDURES

In keeping with the hypothesis that the first quarto (Q) of *The Merchant of Venice* derives from a fair copy of Shakespeare's manuscript, prepared either by him or a theatrical scribe, Q has been used as the control text for this edition, as it has been for other modern editions. In so far as evidence indicates that the Folio version (F), while based upon Q, may have been collated against a prompt-book used in the theatre, stage directions in F not appearing in Q have been adopted. These along with all other divergences from the control text other than mere accidentals are recorded in the collation. Where stage directions may be questioned, broken brackets are used.

Since this is a modern-spelling edition, changes in spelling that reflect modernization only are not recorded in the collation, including changes in punctuation, except where such changes may alter meaning. Within the collation, the authority for the reading adopted in this edition appears immediately after the lemma, followed by other readings in their original spellings (except for long *s*) in chronological order. Where an old-spelling edition for another playwright or author is used in the commentary or elsewhere, spellings are modernized. Speech prefixes have been normalized, but significant alterations, such as the variations of '*Shyl.*' and '*Iewe.*', are recorded in the collation, as these may indicate illuminating attitudinal references. All indications for speeches to be spoken *aside* or to a particular person or persons (e.g. *To Shylock*) are editorial and therefore not listed in the collation. Although iambic pentameter is the basic metrical form Shakespeare used, hexameters occasionally appear, usually as long lines (as at 1.3.122, 2.7.5) but sometimes as half lines that may be joined (as at 4.1.1). The collation records those instances where modern editors prefer different lineation (as at 4.1.316).

Unless otherwise stated, biblical quotations are from the Geneva Bible of 1560, as some evidence indicates that Shakespeare was reading this version in the 1596 edition at the time he was writing this play, although he was also familiar with the Bishops' Bible of 1568. References to Shakespeare's works are keyed to the *Complete Works*, General Editors Stanley Wells and Gary Taylor.

Words are normally defined only when they first appear. Glossed words are listed in the index.

Abbreviations and References

The following abbreviations are used in the introduction, collations, and commentary. The place of publication is, unless otherwise specified, London.

EDITIONS OF SHAKESPEARE

Q	*The most excellent Historie of the Merchant of Venice.* Written by William Shakespeare. 1600
Q2	*The Excellent History of the Merchant of Venice.* Written by W. Shakespeare. 1600 [for 1619]
F	The First Folio, 1623
F2	The Second Folio, 1632
F3	The Third Folio, 1663
F4	The Fourth Folio, 1685
Bevington	David Bevington, *The Merchant of Venice*, Bantam Shakespeare (New York, 1988)
Brown	John Russell Brown, *The Merchant of Venice*, The Arden Shakespeare (1955)
Cambridge	W. G. Clark and W. A. Wright, *Works*, The Cambridge Shakespeare, 9 vols. (Cambridge, 1863–6), vol. ii
Capell	Edward Capell, *Comedies, Histories, and Tragedies*, 10 vols. (1767–8), vol. iii
Collier	John Payne Collier, *Works*, 8 vols. (1842–4), vol. ii
Delius	N. Delius, *Complete Works of William Shakespeare*, 3rd edn. (1872)
Dyce	Alexander Dyce, *Works*, 6 vols. (1857), vol. ii
Eccles	*The Comedy of The Merchant of Venice* (1805)
Furness	Horace Howard Furness, *The Merchant of Venice*, A New Variorum Edition (Philadelphia, 1888)
Halliwell	James O. Halliwell, *Works*, 16 vols. (1856), vol. v
Hanmer	Thomas Hanmer, *Works*, 6 vols. (Oxford, 1743–4), vol. ii
Johnson	Samuel Johnson, *Plays*, 8 vols. (1765), vol. i
Keightley	Thomas Keightley, *Plays*, 6 vols. (1864)

Kittredge	George Lyman Kittredge, *Works*, revised by Irving Ribner (Boston, 1972)
Malone	Edmond Malone, *Plays and Poems*, 10 vols. (1790), vol. v
Merchant	W. Moelwyn Merchant, *The Merchant of Venice*, The New Penguin Shakespeare (Harmondsworth, 1967)
NCS	M. M. Mahood, *The Merchant of Venice*, New Cambridge Shakespeare (Cambridge, 1987)
Neilson and Hill	A. Neilson and C. J. Hill, *Complete Plays and Poems of William Shakespeare* (Boston, Mass., 1942)
NS	Sir Arthur Quiller-Couch and John Dover Wilson, *The Merchant of Venice*, New Shakespeare (Cambridge, 1953)
Oxford	Wells and Taylor (gen. eds.), *Works* (Oxford, 1986): *The Merchant of Venice* was ed. by William Montgomery.
Pooler	*The Merchant of Venice*, ed. C. K. Pooler (1905)
Pope	Alexander Pope, *Works*, 6 vols. (1723–5)
Riverside	G. B. Evans (textual editor), *The Riverside Shakespeare* (Boston, 1974)
Rowe	Nicholas Rowe, *Works*, 6 vols. (1709), vol. ii
Rowe 1714	Nicholas Rowe, *Works*, 8 vols. (1714), vol. ii
Staunton	Howard Staunton, *Plays*, 3 vols. (1858–60), vol. i
Steevens	Samuel Johnson and George Steevens, *Plays*, 10 vols. (1773), vol. iii
Theobald	Lewis Theobald, *Works*, 7 vols. (1733), vol. ii
Thirlby	(unpublished conjectures in marginal notes of his copies of Shakespeare)
Warburton	William Warburton, *Works*, 8 vols. (1747)
Var. 1785	Samuel Johnson and George Steevens, revised by Isaac Reed, *Plays*, 3rd edn., 10 vols. (1785)
Var. 1793	Samuel Johnson and George Steevens, *Plays*, 15 vols. (1793)
Var. 1803	Samuel Johnson and George Steevens, revised by Isaac Reed, *Plays*, 5th edn. (1803)

OTHER WORKS

Abbott	E. A. Abbott, *A Shakespearian Grammar*, 2nd edn. (1870)
Barber	C. L. Barber, *Shakespeare's Festive Comedy* (Princeton, NJ, 1959)
Barton	John Barton, *Playing Shakespeare* (1984)
Brown, 'Realization'	John Russell Brown, 'The Realization of Shylock', in Brown and Bernard Harris (eds.), *Early Shakespeare* (1961), 186–209

Bullough	Geoffrey Bullough, *Narrative and Dramatic Sources of Shakespeare*, 8 vols. (1957–75)
Bulman	James C. Bulman, *Shakespeare in Performance: 'The Merchant of Venice'* (Manchester, 1991)
Cercignani	Fausto Cercignani, *Shakespeare's Works and Elizabethan Pronunciation* (Oxford, 1981)
Colman	E. A. M. Colman, *The Dramatic Use of Bawdy in Shakespeare* (1974)
Danson	Lawrence Danson, *The Harmonies of 'The Merchant of Venice'* (1978)
Dent	R. W. Dent, *Shakespeare's Proverbial Language: An Index* (1981)
Granville-Barker	Harley Granville-Barker, *Prefaces to Shakespeare*, Second Series (1939)
Greg, *SFF*	W. W. Greg, *The Shakespeare First Folio* (Oxford, 1955)
Fischer	Sandra K. Fischer, *Econolingua* (Newark, Del., 1985)
Holland	Norman Holland, *Psychoanalysis and Shakespeare* (New York, 1966)
Jonson	*Ben Jonson*, ed. C. H. Herford and Percy and Evelyn Simpson, 11 vols. (Oxford, 1925–52)
Kökeritz	Helge Kökeritz. *Shakespeare's Pronunciation* (New Haven, Conn., 1953)
Leggatt	Alexander Leggatt, *Shakespeare's Comedy of Love* (1974)
Lewalski	Barbara K. Lewalski, 'Biblical Allusion and Allegory in *The Merchant of Venice*', *SQ* 13 (1962), 327–43
Marlowe	Christopher Marlowe, *Complete Plays*, ed. Irving Ribner (New York, 1963)
McPherson	David C. McPherson, *Shakespeare, Jonson, and the Myth of Venice* (Newark, Del., 1990)
Noble	Richmond Noble, *Shakespeare's Biblical Knowledge* (1935)
Onions	C. T. Onions, *Shakespeare Glossary*, enlarged and revised by Robert D. Eagleson (1986)
Overton	Bill Overton, *Text & Performance: 'The Merchant of Venice'* (Atlantic Highlands, NJ, 1987)
Oz	Avraham Oz, 'The Egall Yoke of Love: Prophetic Unions in *The Merchant of Venice*', *Assaph*, Section C, No. 3 (1986), 75–108
Rubinstein	Frankie Rubinstein, *A Dictionary of Shakespeare's Sexual Puns and their Significance* (1984)
SAB	*Shakespeare Association Bulletin*
SB	*Studies in Bibliography*

Schmidt	Alexander Schmidt, *A Shakespeare Lexicon*, 4th edn. (revised by G. Sarrazin), 2 vols. (Berlin and Leipzig, 1923)
Shaheen	Naheeb Shaheen, *Biblical References in Shakespeare's Comedies* (Newark, Del., 1992)
Sisson	C. J. Sisson, *New Readings in Shakespeare*, 2 vols. (Cambridge, 1956)
SQ	*Shakespeare Quarterly*
SStud	*Shakespeare Studies*
SSur	*Shakespeare Survey*
TC	Stanley Wells and Gary Taylor, with John Jowett and William Montgomery, *William Shakespeare: A Textual Companion* (Oxford, 1987)
Tilley	Morris Palmer Tilley, *A Dictionary of the Proverbs in England in the Sixteenth and Seventeenth Centuries* (Ann Arbor, Mich., 1950)
Walker	W. S. Walker, *A Critical Examination of Shakespeare's Text* (1860)
Wright	George T. Wright, *Shakespeare's Metrical Art* (Berkeley, Calif., 1988)

The Merchant of Venice

THE PERSONS OF THE PLAY

DUKE of Venice

PORTIA, the lady of Belmont

NERISSA, her gentlewoman

PRINCE OF MOROCCO ⎫
PRINCE OF ARRAGON ⎭ Portia's suitors

BALTHASAR ⎫
STEFANO ⎭ Portia's servants

BASSANIO

LEONARDO, Bassanio's servant

ANTONIO, a merchant of Venice, Bassanio's friend

GRAZIANO, Bassanio's friend

LORENZO, Bassanio's friend, in love with Jessica

SALARINO ⎫
SOLARIO ⎬ Venetian gentlemen, Antonio's friends
SALERIO ⎭

SHYLOCK, a Jewish money-lender

JESSICA, his daughter, in love with Lorenzo

LANCELOT GOBBO, Shylock's servant, later Bassanio's

OLD GOBBO, Lancelot's father

TUBAL, another Jewish moneylender

GAOLER

Magnificoes, musicians, officers, and attendants of the court

The Merchant of Venice

1.1   *Enter Antonio, Salarino, and Solanio*

ANTONIO

In sooth, I know not why I am so sad.
It wearies me, you say it wearies you;
But how I caught it, found it, or came by it,
What stuff 'tis made of, whereof it is born,
I am to learn;   5
And such a want-wit sadness makes of me
That I have much ado to know myself.

SALARINO

Your mind is tossing on the ocean,
There where your argosies with portly sail,
Like signors and rich burghers on the flood,   10
Or as it were the pageants of the sea,
Do overpeer the petty traffickers
That curtsy to them, do them reverence,
As they fly by them with their woven wings.

The Merchant of Venice] F; *The comicall History of the Mer-chant of Venice* Q subs.
   1.1] *after* Rowe; *Actus Primus* F; *not in* Q   0.1 *Salarino*] Q, F, NCS; *Salerio* NS, BROWN,
MERCHANT, OXFORD   0.1 *Solanio*] CAPELL; *Salanio* Q, F   5–6] *after* Q3; *as one line* Q, F
13 curtsy] F; cursie Q

1.1 The play opens in Venice and then
shifts to and from Belmont.

0.1 **Salarino** See Collation and Textual
Introduction.

1 **sooth** truth
sad melancholy, morose. Many explanations have been offered by critics, but
not Antonio, who helplessly confesses
his ignorance, 3–5, and rejects his
friends' explanations, 8–50.

6 **want-wit** idiot, fool

7 **I . . . myself** i.e. his melancholy has
made him so absent-minded that he
scarcely knows who he is. An allusion
to the ancient imperative *nosce teipsum*
('know thyself') is likely (see Danson,
40–3).

9 **argosies** large merchant ships
portly stately; but the next line suggests big-bellied corpulence, i.e. wind

in the ships' sails billowing them out

10 **signors** gentlemen
burghers freemen of boroughs, pros-
perous citizens

11 **pageants** pageant wagons. Large, elaborate floats were common in Elizabethan pageantry. NCS cites Alice
Venezky, *Pageantry on the Elizabethan
Stage* (New York, 1951), 102.

12 **overpeer** (1) rise above (2) look over
(and hence down upon)
petty traffickers i.e. smaller vessels

13 **curtsy** The rocking of the smaller ships
in the larger one's wake may suggest
the simile of curtsying, or bowing (Furness). But NCS cites A. F. Falconer,
Shakespeare and the Sea (1965), 22:
small cargo ships lower their topsails
as a mark of respect.

14 **woven wings** i.e. sails

SOLANIO

> Believe me, sir, had I such venture forth,   15
> The better part of my affections would
> Be with my hopes abroad. I should be still
> Plucking the grass to know where sits the wind,
> Peering in maps for ports and piers and roads,
> And every object that might make me fear   20
> Misfortune to my ventures out of doubt
> Would make me sad.

SALARINO   My wind cooling my broth
> Would blow me to an ague when I thought
> What harm a wind too great might do at sea.
> I should not see the sandy hour-glass run   25
> But I should think of shallows and of flats,
> And see my wealthy Andrew docked in sand,
> Vailing her high-top lower than her ribs
> To kiss her burial. Should I go to church
> And see the holy edifice of stone   30
> And not bethink me straight of dangerous rocks
> Which, touching but my gentle vessel's side,
> Would scatter all her spices on the stream,
> Enrobe the roaring waters with my silks,
> And, in a word, but even now worth this,   35

19 Peering] F; Piring Q, BROWN, NCS; Piering Q2; Prying Q3   27 docked] ROWE; docks
Q, F; decks OXFORD, *after* Delius (*conj.* Collier)

15 **venture** risky enterprise, as Shylock
   notes, 1.3.17
16 **affections** feelings, thoughts
17 **still** always
18 **Plucking the grass** Tossing leaves of
   grass in the air is an old way of determining wind direction.
19 **Peering** See Collation. Adopting Q1
   'Piring' (Brown, NCS) does not, in
   Shakespeare's pronunciation, avoid
   the jingle with 'piers' (see Cercignani,
   148), and Q3 'Prying' has no authority. In any case, although etymologically distinct, 'peering' and 'piring'
   have the same meaning.
   roads anchorages, as in *Shrew* 2.1.371
21 **out of doubt** certainly
22 **wind** breath
23 **ague** shivering fever; malaria was
   thought to be caused by bad air
26 **flats** shoals
27 **Andrew** name of a ship. The English

captured the *San Andrés*, or *Andrew*, in
Cadiz harbour in 1596.
docked See Collation. The Q, F misprint, if it is one, probably derives from
a misreading of manuscript 'dockt'.
Collier's conjecture presumes a less
likely misreading, *e/o*. 'Docked in sand'
hardly implies a safe harbourage, as
Walker (cited by *TC*) suggests.
28 **Vailing** lowering
   high-top topmast
29 **To . . . burial** i.e. to salute her death
   in the sand
31 **straight** straightaway
32 **gentle** (1) noble (2) delicate, vulnerable
33–4 **spices . . . silks** typical trading merchandise from the East, sought by
   Europeans. Brown notes that Barabas's
   argosy was so loaded, *Jew of Malta*
   1.1.45.
35 **even now worth this** just now worth
   so much

And now worth nothing? Shall I have the thought
To think on this, and shall I lack the thought
That such a thing bechanced would make me sad?
But tell not me; I know Antonio
Is sad to think upon his merchandise.   40

ANTONIO
Believe me, no. I thank my fortune for it,
My ventures are not in one bottom trusted,
Nor to one place; nor is my whole estate
Upon the fortune of this present year:
Therefore my merchandise makes me not sad.

SOLANIO
Why, then you are in love.

ANTONIO   Fie, fie.

SOLANIO
Not in love neither. Then let us say you are sad
Because you are not merry; and 'twere as easy
For you to laugh, and leap, and say you are merry
Because you are not sad. Now, by two-headed Janus,   50
Nature hath framed strange fellows in her time:
Some that will evermore peep through their eyes
And laugh like parrots at a bagpiper,
And other of such vinegar aspect
That they'll not show their teeth in way of smile,   55
Though Nestor swear the jest be laughable.

36 nothing?] Q2; ~. Q, F   47 neither.] ~ : Q, F; ~ ? Q2

42 **My . . . trusted** Antonio alludes to the
proverb 'Venture not all in one bottom' (Tilley A209); compare *1 Henry
VI* 4.6.32–3: 'O, too much folly is it
. . . | To hazard all our lives in one
small boat.'
bottom keel or hull; hence, ship
(Onions)
47–50 **Then . . . sad** These jesting tautologies, also proverbial, like the rest of
the speech they introduce, are weak
attempts to cheer Antonio up.
49 **laugh, and leap** proverbial signs of joy
(Dent L92a.1, citing Jonson, *Every
Man Out of his Humour* 1.3.120–1: 'To
sit and clap my hands, and laugh and
leap, | Knocking my head against my
roof with joy')
50 **two-headed Janus** The Roman god of
exits and entrances had two faces, not

heads, looking in opposite directions,
one face smiling and the other frowning, thus suggesting opposing attitudes, like comic and tragic masks.
52 **peep . . . eyes** During a hearty laugh,
the eyes appear half shut (Warburton).
53 **laugh . . . bagpiper** The parrot, prover-
bially a foolish bird (as in Cassio's allusion, *Othello* 2.3.273), laughs even
at the melancholy sound of a bagpipe.
54–6 **other . . . laughable** By contrast,
others will not crack a smile even
though gravity, personified by hoary
Nestor, would permit it (Brown).
54 **other** an old plural form (Abbott 12)
aspect accented on the second syllable
56 **Nestor** In Homer's *Iliad*, the oldest and
most venerable of the Greek generals
opposing Troy.

Enter Bassanio, Lorenzo, and Graziano
Here comes Bassanio, your most noble kinsman,
Graziano, and Lorenzo. Fare ye well,
We leave you now with better company.

SALARINO
I would have stayed till I had made you merry,   60
If worthier friends had not prevented me.

ANTONIO
Your worth is very dear in my regard.
I take it your own business calls on you,
And you embrace th'occasion to depart.

SALARINO
Good morrow, my good lords.   65

BASSANIO
Good signors both, when shall we laugh? Say, when?
You grow exceeding strange: must it be so?

SALARINO
We'll make our leisures to attend on yours.
                              Exeunt Salarino and Solanio

LORENZO
My Lord Bassanio, since you have found Antonio,
We two will leave you; but at dinner-time   70
I pray you have in mind where we must meet.

BASSANIO
I will not fail you.

GRAZIANO
You look not well, Signor Antonio;
You have too much respect upon the world.
They lose it that do buy it with much care.   75
Believe me, you are marvellously changed.

57 **kinsman** Perhaps used figuratively, as a blood relationship between Antonio and Bassanio is nowhere else mentioned. In *Il Pecorone* Giannetto is Ansaldo's adopted son.

61 **prevented** forestalled. Why Salarino and Solanio depart here is not entirely clear, although the ensuing dialogue suggests jealousy or rivalry for Antonio's affections, hence a certain coolness or distance (67–8).

64 **occasion** opportunity

66 **when . . . laugh** i.e. get together for some fun

67 **strange** distant, estranged

68 **We'll . . . yours** polite way of saying 'We're ready to get together when you are'

70 **We . . . you** Graziano here fails to follow Lorenzo's cue, that it is tactful to leave Bassanio alone with Antonio (Brown, following Pooler).

74 **respect upon** care, regard for

75 **They . . . care** The real value of an object (here, 'the world', i.e. social and material considerations) is lost when worried about excessively. Compare Matt. 16: 25: 'For whosoever will

ANTONIO

I hold the world but as the world, Graziano,
A stage where every man must play a part,
And mine a sad one.

GRAZIANO   Let me play the fool:
With mirth and laughter let old wrinkles come,   80
And let my liver rather heat with wine
Than my heart cool with mortifying groans.
Why should a man, whose blood is warm within,
Sit like his grandsire cut in alabaster,
Sleep when he wakes, and creep into the jaundice   85
By being peevish? I tell thee what, Antonio—
I love thee, and 'tis my love that speaks—
There are a sort of men whose visages
Do cream and mantle like a standing pond,
And do a wilful stillness entertain,   90
With purpose to be dressed in an opinion

87 'tis] Q; it is F, NCS

save his life, shall lose it: and whosoever shall lose his life for my sake, shall find it.' But NCS says the thought there is different and cites G. C. Rosser (ed.), *Merchant of Venice* (1964) instead: 'Those who take the world too seriously find they have lost the capacity to enjoy it.' Shaheen compares 1 John 2: 15.

78–9 **A stage ... one** The world as a stage was an Elizabethan commonplace (Tilley W882). Compare Jaques's speech, 'All the world's a stage', *As You Like It* 2.7.139–66.

79 **Let . . . fool** Emphasis should fall on 'me' (Kittredge). Graziano contrasts his preferred role with Antonio's accepted one. As often noted, Graziano was the name of the comic doctor in the *commedia dell'arte*. Florio's Italian dictionary defines *Gratiano* as 'a gull, a fool or clownish fellow in a play or comedy' (M. J. Levith, *What's in Shakespeare's Names* (1978), 79; cited by NCS).

80 **old** (1) customary, familiar (2) copious, plentiful, as at 4.2.15; but the phrase also suggests growing old and wrinkled with 'mirth and laughter' instead of worry.

81 **heat with wine** The liver was considered the seat of passions, which wine could inflame.

82 **mortifying** (1) penitential (2) causing death (NCS). Sighs and groans were believed to drain blood from the heart, as in *2 Henry VI* 3.2.60–1.

84 **Sit . . . alabaster** Effigies cut in stone of kneeling figures were familiar in churches. 'Sit' = 'to rest the body on the knees, to be in a kneeling posture' (*OED* 12).

85–6 **creep . . . peevish** Jaundice was believed to result from an excess of yellow bile, which could be caused by emotional disorders. Compare *Troilus* 1.3.1: 'Princes, what grief hath set the jaundice on your cheeks?'

89 **cream and mantle** turn pale and masklike. The imagery derives from cream floating above milk and algae floating on a stagnant (*standing*) pond. Compare 'cream-faced loon', 'whey-face', *Macbeth* 5.3.11, 19, and 'the green mantle of the standing pool', *Lear* 3.4.125.

90 **entertain** maintain

91 **dressed** The metaphor emphasizes insincerity.
    opinion reputation

Of wisdom, gravity, profound conceit,
As who should say, 'I am Sir Oracle,
And when I ope my lips let no dog bark.'
O my Antonio, I do know of these   95
That therefore only are reputed wise
For saying nothing, when I am very sure
If they should speak, would almost damn those ears
Which, hearing them, would call their brothers fools.
I'll tell thee more of this another time.   100
But fish not with this melancholy bait
For this fool gudgeon, this opinion.—
Come, good Lorenzo.—Fare ye well awhile;
I'll end my exhortation after dinner.

LORENZO

Well, we will leave you then till dinner-time.   105
I must be one of these same dumb wise men,
For Graziano never lets me speak.

GRAZIANO

Well, keep me company but two years more,
Thou shalt not know the sound of thine own tongue.

---

93 Sir] POPE; sir Q; sir an F   98 damn] F4; dam QI–2, F, KITTREDGE; damme F2–4
103 Fare ye well] Q, F; farwell Q2

92 **conceit** 'understanding, mental capacity' (Onions), as in *As You Like It* 5.2.52
93 **As . . . say** as if they were to say **Sir Oracle** Compare 'Sir Smile', *Winter's Tale* 1.2.197. The Folio's interpolation of 'an' disrupts both the metre and the joke.
94 **let . . . bark** proverbial (Tilley D526). Compare Exod. 11: 7: 'But against none of the children of Israel shall a dog move his tongue . . . that ye may know that the Lord putteth a difference between the Egyptians and Israel' (NCS).
95–7 **these . . . nothing** Compare Prov. 17: 28: 'Even a fool (when he holdeth his peace) is counted wise, and he that stoppeth his lips, prudent', and Job 13: 5 (Noble).
98–9 **damn . . . fools** i.e. would risk damnation by calling them fools for what they have said ('ears' is synecdochic). Compare Matt. 5: 22: 'And who-

soever shall say, Fool, shall be worthy to be punished with hell fire' (Noble). Kittredge sees a pun, *dam–damn*, and in fact follows QI–2 spelling (see Collation). The primary sense then = 'cause their hearers to stop up their ears against the foolishness they are hearing'.
101–2 **fish . . . opinion** Graziano warns Antonio not to try to win the reputation of wisdom by adopting silent melancholy as a 'bait' or lure for foolish reputation. The gudgeon was notoriously a gullible small fish.
104 **end . . . dinner** 'The humour of this consists in its being an allusion to the practice of the puritan preachers of the times; who, being generally very long and tedious, were often forced to put off that part of their sermon called the *exhortation*, till after dinner' (Warburton, cited by Malone).
106 **dumb wise men** alluding to lines 90–9

ANTONIO

Fare you well. I'll grow a talker for this gear.   110

GRAZIANO

Thanks, i'faith, for silence is only commendable

In a neat's tongue dried and a maid not vendible.

Exeunt Graziano and Lorenzo

ANTONIO   Yet is that anything now?

BASSIANO   Graziano speaks an infinite deal of nothing,

more than any man in all Venice. His reasons are as   115

two grains of wheat hid in two bushels of chaff: you

shall seek all day ere you find them, and when you

have them, they are not worth the search.

ANTONIO

Well, tell me now what lady is the same

To whom you swore a secret pilgrimage,   120

That you today promised to tell me of.

BASSANIO

'Tis not unknown to you, Antonio,

How much I have disabled mine estate

By something showing a more swelling port

Than my faint means would grant continuance,   125

Nor do I now make moan to be abridged

From such a noble rate; but my chief care

Is to come fairly off from the great debts

---

110 Fare you well] Q, F; Farwell Q2   112 tongue] Q2, F; togue Q   112.1 *Exeunt*] Q;
Exit F   *Graziano and Lorenzo*] THEOBALD *subs*.; *not in* Q, F   113 Yet is] OXFORD; It is Q,
F, NCS, BROWN; Is ROWE +   now?] ROWE, OXFORD; ~. Q, F, NCS, BROWN; new (*conj.*
Johnson)   114 nothing,] Q2, F; ~_∧_ Q   121 of.] Q; of? F

---

110 **gear** discourse, talk (Onions)
112 **neat's tongue dried** cured ox tongue,
    a delicacy, with a bawdy allusion to
    an impotent old man. Compare Field,
    A Woman is a Weathercock (1612),
    1.2: 'But did that little old dried neat's
    tongue, that eel-skin, [be]get him?'
    (Brown).
112 **vendible** saleable, i.e. marriageable
    (Onions)
113 **Yet . . . now?** See Collation. 'The
    compositor may have misread a hastily
    written "yet" as "yt" ' (Wells, *TC* 324).
120 **secret pilgrimage** Religious imagery
    was commonplace for lovers; compare
    2.7.40. But Bassanio's trip to Belmont
    is hardly a secret: everyone seems to

know where he is going. The reference
may be a vestige from *Il Pecorone*,
where Giannetto carefully concealed
his quest from everyone (NCS); on the
other hand, only after gaining Antonio's support would Bassanio have
confided in his friends.
124 **something** somewhat, to some extent
    (as at 129)
    swelling port 'expensive equipage and
    external pomp of appearance' (Malone)
126–7 **Nor . . . rate** i.e. nor do I now
    complain ('make moan') about being
    curtailed from that affluent style
    ('noble rate')
128 **come fairly off** properly, honourably
    extricate myself

Wherein my time something too prodigal
Hath left me gaged. To you, Antonio,   130
I owe the most in money and in love,
And from your love I have a warranty
To unburden all my plots and purposes
How to get clear of all the debts I owe.

ANTONIO

I pray you, good Bassanio, let me know it,   135
And if it stand, as you yourself still do,
Within the eye of honour, be assured
My purse, my person, my extremest means,
Lie all unlocked to your occasions.

BASSANIO

In my schooldays, when I had lost one shaft,   140
I shot his fellow of the selfsame flight
The selfsame way with more advisèd watch
To find the other forth, and by adventuring both
I oft found both. I urge this childhood proof
Because what follows is pure innocence.   145
I owe you much, and, like a wilful youth,
That which I owe is lost; but if you please
To shoot another arrow that self way
Which you did shoot the first, I do not doubt,
As I will watch the aim, or to find both   150
Or bring your latter hazard back again
And thankfully rest debtor for the first.

---

130 **gaged** entangled (Onions); engaged,
bound (Schmidt)

132 **warranty** authorization

135, 136 **it . . . it** i.e. the means to 'get
clear' of his debts

136-7 **if . . . honour** i.e. if your plan can
be looked upon as honourable, as you
yourself always are (despite your
spendthrift ways)

139 **occasions** needs, requirements

140-4 **In . . . both** Shooting a second
arrow to find a lost first one was
proverbial (Tilley A325). Cf. Robert
Armin, *Quips upon Questions* (1600),
D1: 'Another shaft they shoot that
direct way | As whilom they the first
shot, . . . | . . . [So] The former arrow
may be found again' (Collier, cited by

Brown).

140 **shaft** arrow

141 **his** its (as often)
flight Arrows of the same size and
weight would have the same power of
flight (Onions).

142 **advisèd watch** careful observation

143 **adventuring** risking

144 **proof** 'truth or knowledge gathered
by experience' (Schmidt). Compare
Much Ado 2.1.171: 'This is an accident of hourly proof.'

145 **innocence** ingenuousness (NCS)

148 **self** i.e. selfsame

150-1 **or . . . Or** either . . . Or

151 **hazard** stake. 'A key word in the
play, linking the choice of caskets with
Antonio's risks' (NCS).

ANTONIO

You know me well, and herein spend but time
To wind about my love with circumstance;
And out of doubt you do me now more wrong
In making question of my uttermost
Than if you had made waste of all I have.
Then do but say to me what I should do
That in your knowledge may by me be done,
And I am pressed unto it: therefore speak.

BASSANIO

In Belmont is a lady richly left,
And she is fair and, fairer than that word,
Of wondrous virtues. Sometimes from her eyes
I did receive fair speechless messages.
Her name is Portia, nothing undervalued   165
To Cato's daughter, Brutus' Portia;
Nor is the wide world ignorant of her worth,
For the four winds blow in from every coast
Renownèd suitors, and her sunny locks
Hang on her temples like a golden fleece,   170
Which makes her seat of Belmont Colchis' strand,
And many Jasons come in quest of her.
O my Antonio, had I but the means
To hold a rival place with one of them,
I have a mind presages me such thrift   175
That I should questionless be fortunate.

---

155 me now] Q; *not in* F   171 Colchis'] RIVERSIDE; Cholchos Q, F; Colchos Q2

153 **spend but time** only waste time
154 **To . . . circumstance** i.e. to approach me and my affection for you with circuitous discourse
156 **In . . . uttermost** in questioning that I will do all I can
160 **pressed** impelled. Many editions (e.g. Brown) retain the Q/F spelling 'prest' to suggest derivation from OF *prest*, modern *pret* ('ready'); but the construction with 'unto' indicates a verb form.
161 **richly left** i.e. bequeathed a large fortune
163 **Sometimes** once, on a particular occasion (Onions)

166 **Cato's . . . Portia** Shakespeare was soon to dramatize Portia's nobility as Brutus's wife in *Julius Caesar*. Her father was the noble tribune Cato Uticensis. See *Caesar* 2.1.294–6.
169–72 **her sunny . . . of her** Bassanio compares Portia's fair hair to the highly prized golden fleece which, in Greek legend, after many adventures Jason and his Argonauts brought back from Colchis on the Black Sea. Cf. 3.2.239.
171 **strand** shore
175 **presages** that augurs
thrift (1) success (2) profit (as at 1.3.47)

ANTONIO

Thou know'st that all my fortunes are at sea,

Neither have I money nor commodity

To raise a present sum. Therefore go forth—

Try what my credit can in Venice do;

That shall be racked, even to the uttermost,   180

To furnish thee to Belmont, to fair Portia.

Go presently inquire, and so will I,

Where money is, and I not question make

To have it of my trust or for my sake.   *Exeunt* 185

1.2   *Enter Portia with her waiting woman, Nerissa*

PORTIA By my troth, Nerissa, my little body is aweary of this great world.

NERISSA You would be, sweet madam, if your miseries were in the same abundance as your good fortunes are; and yet, for aught I see, they are as sick that surfeit with too much as they that starve with nothing. It is no mean happiness, therefore, to be seated in the mean: superfluity comes sooner by white hairs, but competency lives longer.

PORTIA Good sentences and well pronounced.

NERISSA They would be better if well followed.

PORTIA If to do were as easy as to know what were good

1.2] *after* Rowe; *not in* Q, F   7 mean happiness] Q; smal ~ F

176 **That . . . fortunate** Brown believes
Bassanio does not know about the caskets, since his earlier visit to Belmont
occurred when Portia's father still
lived (see 1.2.109–18). But if Bassanio
knows about the suitors, he may also
know about the test. Nevertheless, he
believes he is favoured (163–4).

178 **commodity** goods, merchandise

179 **present sum** ready amount of cash

181 **racked** stretched, strained

183 **presently** at once (as often)

185 **of . . . sake** i.e. either on my credit
or for friendship's sake (NS)

1.2.0 The conversation apparently takes
place in the casket room. See 29,
below.

1 **troth** faith
aweary Some commentators (e.g. Merchant) find an echo here of Antonio's
opening melancholy, but Portia's

'weariness' is not Antonio's 'sadness'.
It derives from ennui and is immediately mocked by Nerissa.

7–8 **mean . . . mean** Nerissa plays on
two senses of 'mean': (1) contempt-
ible, (2) midpoint, and alludes to the
proverbial *via media* (Tilley V80). F's
alteration destroys the word-play.

8–9 **superfluity . . . longer** i.e. excess ages
us faster than moderation, which fos-
ters longer life

9 **competency** sufficiency

10 **sentences** *sententiae*, maxims
pronounced delivered

12–15 **If . . . instructions** Portia, too, responds with proverbial wisdom, alluding to the old sayings 'If wishes were
horses beggars would ride' and 'Practise what you preach' (Tilley P537a)
(NCS). Shaheen compares Rom. 2:
21: 'Thou therefore which teachest

to do, chapels had been churches and poor men's cottages princes' places. It is a good divine that follows his own instructions. I can easier teach twenty what were good to be done than to be one of the twenty to follow mine own teaching. The brain may devise laws for the blood, but a hot temper leaps o'er a cold decree: such a hare is madness, the youth, to skip o'er the meshes of good counsel, the cripple. But this reasoning is not in the fashion to choose me a husband. O me, the word 'choose'! I may neither choose who I would, nor refuse who I dislike; so is the will of a living daughter curbed by the will of a dead father. Is it not hard, Nerissa, that I cannot choose one nor refuse none?

NERISSA Your father was ever virtuous, and holy men at their death have good inspirations; therefore the lott'ry that he hath devised in these three chests of gold, silver, and lead, whereof who chooses his meaning chooses you, will no doubt never be chosen by any rightly but one who you shall rightly love. But what warmth is there in your affection towards any of these princely suitors that are already come?

PORTIA I pray thee overname them; and as thou namest

15
20
25
30
35

16 to be one] Q; ‿ ~ ~ F   20 reasoning] Q; reason F   21 the] Q; *not in* F   22–3 who ... who] Q; whom ... whom F   24 Is it] Q; it is F   29 lott'ry] Q (lottrie); lotterie Q2, F   31 you, will no doubt] Q, F; ~, no doubt you wil Q2   32 you] Q, F; *not in* Q2

another, teachest thou not thyself?'

13 **chapels** small oratories
14 **divine** clergyman
18 **blood** seat of emotion; opposed to reason
  hot temper passionate disposition. A 'hot temper' was the result of an imbalance in the four humours, or fluids, in the body; blood was characteristically a hot humour. In the little allegory that follows, youth is compared to a mad March hare leaping over nets, or snares—the restraints set by good counsel, imagined here as a halting cripple. Compare 'Youth will have its course' (Tilley Y48).
22–3 **who ... who** whom, as at 32. 'The inflection of *who* is frequently neg-

lected' (Abbott 274).
23–4 **will ... will** Portia plays on the senses (1) desire, (2) testament, (3) wilfulness.
27–8 **holy ... inspirations** Brown compares Gaunt's similar proverbial thinking in *Richard II* 2.1.5–11, 31.
29 **these** Nerissa may here point towards the caskets, apparently sizeable and in view from the start of the scene (NCS).
30 **his** its
31 **will no doubt** See Collation. In altering the word order and adding 'you', Q2 changes both the syntax and the sense.
32 **rightly ... rightly** Nerissa puns on (1) correctly, (2) truly (Schmidt).
35 **overname** name from beginning to end. The dialogue that follows recalls *Two Gentlemen* 1.2.4–33, although

them, I will describe them and, according to my description, level at my affection.

NERISSA First there is the Neapolitan prince.

PORTIA Ay, that's a colt indeed, for he doth nothing but   40
talk of his horse, and he makes it a great appropriation
to his own good parts that he can shoe him himself. I
am much afeard my lady his mother played false with
a smith.

NERISSA Then is there the County Palatine.

PORTIA He doth nothing but frown, as who should say   45
'An you will not have me, choose'. He hears merry
tales and smiles not; I fear he will prove the weeping
philosopher when he grows old, being so full of unmannerly sadness in his youth. I had rather be married to a death's-head with a bone in his mouth than to   50
either of these. God defend me from these two!

NERISSA How say you by the French lord, Monsieur le
Bon?

PORTIA God made him, and therefore let him pass for a
man. In truth, I know it is a sin to be a mocker, but   55

---

42 afeard] Q; afraid F   44 Then] Q2; Than Q, F   is there] Q, F; there is Q2   Palatine]
Q2; Palentine Q, F   49 be] Q; to ~ F   53 Bon] CAPELL; Boune Q, F

---

there not Julia but her waiting-woman describes the suitors. Perhaps the success of that scene occasioned this one (NCS).

37 **level at** point to (*OED* 2.6*d*). Presumably the verb is not imperative but forms a compound predicate. Nerissa does not guess at Portia's feelings, which are obvious from her descriptions.

38 **Neapolitan** Neapolitans in Shakespeare's day were famous for horsemanship (Steevens). Nerissa's catalogue is replete with national stereotypes. Portia's comments on her suitors emphasize their foreignness (Overton).

39 **colt** 'witless, heady, gay youngster' (Johnson)

41 **parts** abilities, capacities, as in *As You Like It* 1.1.135: 'an envious emulator of every man's good parts.'

43 **smith** blacksmith

44 **County** Count, as in *Romeo* 5.3.173, *Twelfth Night* 1.5.291. The final syllable apparently derives from French *counte* or Italian *conte* (*OED*).

Palatine Many regions in Europe, including England, had palatinates, or districts having royal privileges. Since a German is later satirically described, the palatinate of the Rhine is not referred to, although Johnson thought a Polish county palatine, Alberto a Lasco, might be. He visited London in 1583, and the dour disposition suggests a middle European. But probably no specific person is meant.

45 **as who** as if one

46 **An** if (as often)
choose i.e. have it your own way (Brown, citing Pooler)

47–8 **weeping philosopher** Heraclitus of Ephesus, who wept at people's errant behaviour. In *Satire X* Juvenal contrasted him with Democritus, the laughing philosopher (Merchant).

48–9 **unmannerly** (1) impolite, (2) unbecoming (to his youth) (NCS)

50 **death's-head . . . mouth** Compare the skull and crossed bones carved on tombstones.

52 **by** concerning (Abbott 145)

These Pages
rous Sterotypes
e explored

I like would
e been how
passes Saw
various
ethnicities

he!—why, he hath a horse better than the Neapolitan's, a better bad habit of frowning than the Count
Palatine. He is every man in no man. If a throstle sing,
he falls straight a-cap'ring. He will fence with his own
shadow. If I should marry him, I should marry twenty   60
husbands. If he would despise me, I would forgive him,
for if he love me to madness, I shall never requite him.

NERISSA What say you then to Falconbridge, the young
baron of England?

PORTIA You know I say nothing to him, for he under-   65
stands not me, nor I him: he hath neither Latin,
French, nor Italian, and you will come into the court
and swear that I have a poor pennyworth in the English. He is a proper man's picture, but alas, who can
converse with a dumb show? How oddly he is suited!   70
I think he bought his doublet in Italy, his round hose
in France, his bonnet in Germany, and his behaviour
everywhere.

NERISSA What think you of the Scottish lord, his neighbour?   75

PORTIA That he hath a neighbourly charity in him, for
he borrowed a box of the ear of the Englishman and

58 Palatine] Q2; Palentine Q, F   throstle] POPE; Trassell Q, F; Tassell F3   62 shall]
Q; should F   74 Scottish] Q; other F

58 **He . . . no man** i.e. lacking an identity
   of his own, he emulates everyone else
58–9 **If . . . cap'ring** 'i.e. he will dance
   whoever calls the tune' (Brown)
   throstle thrush. Q/F 'Trassell' is probably a phonetic or dialectic spelling, if
   not an *o/a* misreading.
61 **If** even if
63 **Falconbridge** Compare Shakespeare's
   hero (a plain blunt Englishman) in
   King John.
67–8 **come . . . swear** i.e. testify; the court
   referred to is judicial, not royal.
69 **proper man's picture** i.e. image of a
   handsome man. Furness believes with
   Allen that 'man's picture', a compound term, should be hyphenated,
   and that 'proper' = 'very'.
70 **dumb show** pantomime, as in *Hamlet*
   3.2.129 SD
   suited dressed, with a possible pun on
   dressed in a suitable way (Brown).
   English eclecticism was a stock joke.

NS cites an analogous passage from
Greene's *Farewell to Folly* (1591) and
Brown cites one from Nashe's *The Un-
fortunate Traveller* (1594).

71 **round hose** It was fashionable to pad
   out stockings to give one a good 'leg'.
74 **Scottish** See Collation. F's 'other' suggests a diplomatic alteration in deference to James I, at whose court the
   play was performed twice in 1605.
76–80 **That he . . . another** Skirmishes
   between Scots and English occurred
   frequently over the centuries; France
   usually sided with Scotland against
   England. Brown notes disorders along
   the border in 1596, resulting in pro-
   clamations by both Elizabeth and
   James to keep the peace. As for 'neighbourly charity', compare Rom. 13:
   10: 'Charity worketh no ill to his
   neighbour' (Bishops' Bible, cited by
   Noble, Shaheen).

swore he would pay him again when he was able. I
think the Frenchman became his surety and sealed
under for another.   80

NERISSA How like you the young German, the Duke of
Saxony's nephew?

PORTIA Very vilely in the morning when he is sober, and
most vilely in the afternoon when he is drunk. When
he is best, he is a little worse than a man, and when he   85
is worst, he is little better than a beast. An the worst
fall that ever fell, I hope I shall make shift to go without him.

NERISSA If he should offer to choose, and choose the
right casket, you should refuse to perform your   90
father's will if you should refuse to accept him.

PORTIA Therefore, for fear of the worst, I pray thee set a
deep glass of Rhenish wine on the contrary casket; for
if the devil be within and that temptation without, I
know he will choose it. I will do anything, Nerissa, ere   95
I will be married to a sponge.

NERISSA You need not fear, lady, the having any of these
lords: they have acquainted me with their determina-
tions, which is indeed to return to their home and to
trouble you with no more suit, unless you may be won   100
by some other sort than your father's imposition depending on the caskets.

PORTIA If I live to be as old as Sibylla, I will die as chaste
as Diana unless I be obtained by the manner of my
father's will. I am glad this parcel of wooers are so   105
reasonable, for there is not one among them but I dote

---

78 **pay him again** pay him back
79 **surety** 'One who makes himself liable
   for the default of another, bail'
   (Onions)
79–80 **sealed under** set his seal to. Note
   Portia's use of legal terminology relat-
   ing to bonds and sureties (Merchant).
80 **for another** i.e. another box of the ear
85–6 **best . . . beast** Kökeritz, 95, hears a
   pun here and in *Dream* 5.1.225–6;
   but Cercignani, 168, claims the word-
   play in both cases depends on antithesis, not identity of sound.
86–7 **An . . . fell** i.e. if the worst come to
   the worst (Tilley W911)

87 **make shift** contrive (Onions)
91 **should refuse** would refuse
93 **Rhenish** i.e. white German wine. See
   3.1.38 n.
   contrary wrong
94 **sponge** The modern slang meaning,
   'one who sponges off someone', is not
   present.
103 **Sibylla** The prophetess of Cumae. Her
   lover, Apollo, granted her as many
   years of life as the grains of sand she
   could hold in one hand. See Ovid's
   Metamorphoses 14.129–53.
104 **Diana** goddess of chastity

on his very absence, and I pray God grant them a fair
departure.

NERISSA Do you not remember, lady, in your father's
time a Venetian, a scholar and a soldier, that came   110
hither in company of the Marquis of Montferrat?

PORTIA Yes, yes, it was Bassanio—as I think so was he
called.

NERISSA True, madam. He of all the men that ever my
foolish eyes looked upon was the best deserving a fair   115
lady.

PORTIA I remember him well, and I remember him
worthy of thy praise.

        Enter a Servingman

How now! What news?

SERVINGMAN The four strangers seek for you, madam, to   120
take their leave, and there is a forerunner come from a
fifth, the Prince of Morocco, who brings word the
Prince his master will be here tonight.

PORTIA If I could bid the fifth welcome with so good
heart as I can bid the other four farewell, I should be   125
glad of his approach. If he have the condition of a saint
and the complexion of a devil, I had rather he should
shrive me than wive me.

Come, Nerissa. (*To the Servingman*) Sirrah, go before.
Whiles we shut the gate upon one wooer,   130
Another knocks at the door.   *Exeunt*

107 pray . . . grant] Q; wish F   112 Bassanio—] RIVERSIDE, *after* Q, F (~,)   think‸]
Q; thinke, F   so . . . he Q, F; he was so Q2   118.1] STAUNTON; *after l.* 119 Q   119
How . . . news] Q; *not in* F   120 for] Q; *not in* F   124 good] Q, F; ~ a Q2   129-31
Come . . . door] OXFORD *lineation; as prose* Q, F

107 **pray . . . grant** See Collation. F's
  alteration reflects the 1606 edict
  against profanity in plays (NCS).
110 **scholar . . . soldier** Compare *Hamlet*
  3.1.154-7 for Ophelia's related allusion to the Renaissance ideal of the
  complete gentleman.
111 **Marquis of Montferrat** Although the
  house of Monferrato had declined by
  the sixteenth century, it had a long
  and distinguished history stretching
  back to the tenth century and through
  the period of the Crusades (Merchant).
112 **as I think** Portia evidently tries to
  cover up her eagerness (Pooler, cited
  by Brown).

119 **How . . . news** F's omission of this
  line may be an oversight or a deliberate cut of inessential matter.
120 **four** Actually, six suitors have been
  described. The discrepancy may be an
  oversight, characteristic of swift composition, and need not reflect revision.
  strangers foreigners
126 **condition** character, disposition. Compare 'A light condition in a beauty
  dark', *L.L.L.* 5.2.20.
127 **complexion . . . devil** Devils traditionally were or wore black. Compare
  Hamlet 4.5.129: 'Vows to the blackest
  devil!'
128 **shrive me** hear my confession and

1.3   *Enter Bassanio with Shylock the Jew*

SHYLOCK   Three thousand ducats. Well.

BASSANIO   Ay, sir, for three months.

SHYLOCK   For three months. Well.

BASSANIO   For the which, as I told you, Antonio shall be
bound.   5

SHYLOCK   Antonio shall become bound. Well.

BASSANIO   May you stead me? Will you pleasure me?
Shall I know your answer?

SHYLOCK   Three thousand ducats for three months and
Antonio bound.   10

BASSANIO   Your answer to that.

SHYLOCK   Antonio is a good man.

BASSANIO   Have you heard any imputation to the contrary?

SHYLOCK   Ho, no, no, no, no. My meaning in saying he is   15
a good man is to have you understand me that he is
sufficient. Yet his means are in supposition. He hath

1.3] *after* Rowe; *not in* Q, F   4] POPE; *as two lines breaking after* 'you' Q, F   7–8] POPE;
as two lines breaking after 'pleasure me' Q, F   9–10] POPE; *as two lines breaking after*
'months' Q, F

give absolution. Portia here and later
(see 2.7.79) expresses racial prejudice.

1.3 The scene returns to Venice.
  0.1 **Shylock the Jew** On Shylock's name,
    see p. 23 above. In the quartos and
    Folio speech headings tend to alternate
    inconsistently between 'Shylock' (ab-
    breviated) and 'Jew', with the latter
    more prominent in Q1 and F than in
    Q2. See Textual Introduction, above.
    For the subtextual implications of the
    ensuing dialogue, see John Russell
    Brown, 'Realization', 196–201.
  1 **Three thousand ducats** The Italian
    gold ducat, worth about 9 shillings,
    was first issued in 1284 and derived
    from the silver Apulian *ducat*, issued in
    1140. The obverse shows the doge of
    Venice kneeling before St Mark; the
    reverse of some coins pictures Christ
    with the inscription 'SIT TIBI XRE DAT
    Q TV REGIS ISTE DVCAT' ('To thee, O
    Christ, be dedicated this duchy which
    you rule'). The name derives from the
    earlier silver ducat and not the last
    word of the inscription, but the figure
    and the inscription are probably why

Shylock refers to them as his 'Christian
ducats' (2.8.16). Two thousand ducats
was the price of a diamond or, per
annum, a good marriage settlement;
3,000 ducats was a good yearly income. See Fischer, 69–70.
  5 **bound** obliged to repay, with a sinister
    implication of 'captive' (NCS)
  7 **stead** assist, help
    pleasure gratify
  9–10 **Three . . . bound** Shylock's repeti-
    tion in measured accents shows a de-
    liberate manner. Bassanio may be
    impatient to close the deal, but Shy-
    lock is in no hurry: he enjoys his
    advantage.
  17 **sufficient** solvent (Onions)
    in supposition uncertain, doubtful—for
    the reasons Shylock gives. Antonio's
    far-flung ventures are indeed risky,
    but their extent indicates Antonio's
    standing as a Venetian tycoon. On
    Antonio's counterparts in Renaissance
    Italy, see Benjamin N. Nelson, 'The
    Usurer and the Merchant Prince: Ita-
    lian Businessmen and the Ecclesiast-
    ical Law of Restitution, 1100–1550',
    Journal of Economic History (Suppl.
    1947), 117.

an argosy bound to Tripolis, another to the Indies. I
understand, moreover, upon the Rialto, he hath a
third at Mexico, a fourth for England, and other ven-   20
tures he hath squandered abroad. But ships are but
boards, sailors but men. There be land rats and water
rats, water thieves and land thieves—I mean pirates—
and then there is the peril of waters, winds, and rocks.
The man is, notwithstanding, sufficient. Three thou-   25
sand ducats; I think I may take his bond.

BASSANIO Be assured you may.

SHYLOCK I will be assured I may, and that I may be assured, I will bethink me. May I speak with Antonio?

BASSANIO If it please you to dine with us.   30

SHYLOCK (*aside*) Yes, to smell pork; to eat of the habita-
tion which your prophet the Nazarite conjured the
devil into. I will buy with you, sell with you, talk with
you, walk with you, and so following; but I will not
eat with you, drink with you, nor pray with you.—   35
What news on the Rialto? Who is he comes here?

 Enter Antonio

BASSANIO This is Signor Antonio.

 (*Bassanio and Antonio speak apart to one another*)

28, 31, 38 SHYLOCK] Q2; *lew.* Q, F   37.1] OXFORD *subs.*; *not in* Q, F

19 **Rialto** Shylock may refer to either the
bridge or the exchange, where Venetian gentlemen and merchants met to
discuss business (compare Coryat,
Crudities (1611), 169, quoted by Furness).

21 **squandered** scattered recklessly (Onions)

23 **pirates** probably a forced pun, although piracy was indeed a real and
terrible danger

24 **peril of waters** Compare 1 Cor. 11:
26: 'I was often in perils of waters, in
perils of robbers', etc. (Shaheen).

27–8 **assured . . . assured** Bassanio means
'reassured', but Shylock takes him in
the sense of requiring financial assur-
ance, or surety. On the word-play here
and its sinister implications, see Danson, 151.

29 **I will bethink me** Shylock may already
be thinking of some ingenious security
he will demand (NCS).

31–5 **Yes . . . you** Probably an aside and
so indicated here, following NS and

Oxford. Open hostility to Bassanio or,
later, Antonio would be inappropriate
at this point (NS). On the other hand,
although he later goes to dine with the
Christians (2.5.11–39), Shylock may
not be able to disguise his aversion to
unkosher food and all it represents to
him, an observant Jew.

31–3 **to eat . . . into** Shylock refers to the
exorcism related in Matt. 8: 28–34.
Christ drove the devils that possessed
two people into a herd of swine.

32 **Nazarite** Actually, Jesus was a Nazar-
ene, but the term only came into use
with the King James Bible (1611).
Until then 'Nazarite' designated anyone from Nazareth as well as anyone
from the Jewish sect that included
Samson (Furness).

37 **This . . . Antonio** 'Not a formal introduction, but simply "It's" ' (NCS).
Approaching Bassanio, Antonio ig-
nores Shylock, who moves downstage
in the pause following Bassanio's

SHYLOCK (*aside*)

How like a fawning publican he looks.
I hate him for he is a Christian,
But more for that in low simplicity
He lends out money gratis and brings down
The rate of usance here with us in Venice.   40
If I can catch him once upon the hip,
I will feed fat the ancient grudge I bear him.
He hates our sacred nation, and he rails,   45
Even there where merchants most do congregate,
On me, my bargains, and my well-won thrift,
Which he calls interest. Cursèd be my tribe
If I forgive him.

BASSANIO   Shylock, do you hear?

SHYLOCK

I am debating of my present store,   50
And by the near guess of my memory
I cannot instantly raise up the gross
Of full three thousand ducats. What of that?
Tubal, a wealthy Hebrew of my tribe,
Will furnish me. But soft! How many months   55

47 well-won] Q (well-wone); well-worne F

half-line to deliver his long aside, really a soliloquy (Overton).

38 **fawning publican** Probably an allusion to the humble publican, or tax collector, who prays in contrast with the self-righteous Pharisee (Luke 18: 9–14), with a reference as well to obsequious innkeepers. Antonio may enter sad and downcast (see 1.1.1), occasioning Shylock's description (NS). Compare the publicans in Matt. 5: 44–7 (Lewalski).

39–44 **I . . . bear him** These lines are often cut in productions that present Shylock as a tragic or sympathetic character. On Shylock's 'cannibalism', see 2.5.14 n.

40 **low simplicity** humble foolishness, perhaps suggesting despicable naïvety, too

41 **gratis** Antonio, unlike Shylock, does not charge interest.

42 **usance** usury, interest, as at 105, 137

43 **upon the hip** at a disadvantage (a wrestling metaphor), used again at

4.1.330

46 **there** presumably, on the Rialto

47 **thrift** Shylock's euphemism for gains earned through usury

48 **tribe** Although 'tribe' could refer to 'each of the twelve divisions of the people of Israel, claiming descent from the twelve sons of Jacob' (*OED* 1), Shylock probably means the Jewish people as a whole, as at 107 below.

49 **Shylock . . . hear** Finished speaking with Antonio, Bassanio breaks into Shylock's meditation.

50 **debating** considering
   present store supply of ready cash

52 **gross** total

54 **Tubal** On his name, see p. 22 above. By involving him in the deal, Shakespeare shows Venetian Jews following the injunction of Deut. 23: 20 to lend money freely to each other and charge interest only to non-Jews (NCS).

55 **soft** i.e. go soft, stop a moment, as in *Hamlet* 5.1.212, *Antony* 2.2.88, etc.

Do you desire? (*To Antonio*) Rest you fair, good signor,
Your worship was the last man in our mouths.

ANTONIO
　Shylock, albeit I neither lend nor borrow
　By taking nor by giving of excess,
　Yet to supply the ripe wants of my friend
　I'll break a custom. (*To Bassanio*) Is he yet possessed
　How much ye would?

SHYLOCK　　　　　　　Ay, ay, three thousand ducats.

ANTONIO　And for three months.

SHYLOCK
　I had forgot; three months; (*to Bassanio*) you told me so.
　Well then, your bond; and let me see. But hear you:　65
　Methoughts you said you neither lend nor borrow
　Upon advantage.

ANTONIO　　　　　　I do never use it.

SHYLOCK
　When Jacob grazed his uncle Laban's sheep—
　This Jacob from our holy Abram was,
　As his wise mother wrought in his behalf,　　　70
　The third possessor; ay, he was the third—

---

61 *To Bassanio*] STAUNTON (*after* 'would', *l.* 62); *not in* Q, F　Is . . . possessed] Q, F; are you resolu'd Q2　62 ye would] Q; he would haue Q2; he ~ F　66 Methoughts] Q, F; Me-thought Q2　71 third—] DYCE *subs.*; ~. Q, F

57 **Your . . . mouths** i.e. we were just speaking of you
　worship 'title of honour given to persons of respectable character, mostly used by inferior persons in addressing their betters' (Schmidt)
59 **excess** Compare Philip Caesar, *A General Discourse against the Damnable Sect of Usurers* (trans. 1578): 'Usury, or as the word of God doth call it, excess . . .' (cited by Brown).
60 **ripe** 'requiring immediate satisfaction' (Onions)
61–2 **Is . . . would** The question is addressed to Antonio, not Shylock, as Q2 and F differently (and mistakenly) emend them. But Shylock's response may have misled the editors (see Collation, and cf. NCS).
61 **possessed** informed of, acquainted with
63 **And . . . months** The short line may indicate a pause, as Shylock should not appear too eager (Brown).
67 **advantage** i.e. interest

68–85 **When Jacob . . . Jacob's** Gen. 27 recounts the story of Rebecca's deception of Isaac so that Jacob, not Esau, became his heir. Thereupon Jacob fled from his brother's anger to his uncle Laban, whom he served for many years, acquiring first Leah, then Rachel, as wives. When the time came for him to leave (Gen. 30: 25–43), Laban agreed to let Jacob take all his particoloured sheep and goats. By placing spotted rods before the eyes of the animals in heat, Jacob encouraged an increase in particoloured offspring, which were his, not Laban's, according to their agreement, and so he prospered. Shylock uses the story to justify his gains through interest: see 91–3, below.
69 **Abram** The first patriarch, Abraham, whose son was Isaac. For metrical reasons, Shakespeare uses Abraham's shorter (and original) name.
71 **third possessor** i.e. of the birthright,

ANTONIO

And what of him? Did he take interest?

SHYLOCK

No, not take interest—not, as you would say,
Directly int'rest. Mark what Jacob did:
When Laban and himself were compromised
That all the eanlings which were streaked and pied
Should fall as Jacob's hire, the ewes being rank
In end of autumn turnèd to the rams,
And when the work of generation was
Between these woolly breeders in the act,   80
The skilful shepherd peeled me certain wands,
And in the doing of the deed of kind
He stuck them up before the fulsome ewes,
Who then conceiving did in eaning time
Fall parti-coloured lambs, and those were Jacob's.   85
This was a way to thrive, and he was blest;
And thrift is blessing, if men steal it not.

ANTONIO

This was a venture, sir, that Jacob served for,
A thing not in his power to bring to pass
But swayed and fashioned by the hand of heaven.   90

78 end] Q, F; th'end Q2
81 peeled] MERCHANT, *after* Pope; pyld Q; pil'd F; pilled BROWN, *following* Knight

as Abraham's grandson. Actors err
in thinking Shylock hesitates here
through uncertainty; on the contrary,
he exults, emphasizing his ancestry,
his descent from Jacob (see Furness).

75 **compromised** agreed
76 **eanlings** new-born lambs
77 **hire** wages
   rank in heat
79 **work of generation** breeding
81 **peeled** collateral form of 'pilled', becoming dominant in this sense in the
   seventeenth century. Q1–2 'pyled' reflects the alternative form used in
   many Tudor Bibles.
   me ethical dative, as in *1 Henry IV*
   4.3.77 (Abbott 220)
82 **deed of kind** act of nature, i.e. mating
83 **fulsome** pregnant
84 **eaning time** time of giving birth
85 **Fall** Drop

86–7 **This . . . blessing** Shylock's justification derives from Gen. 30: 33, which
   the Geneva Bible glosses: 'God shall
   testify for my righteous dealing by rewarding my labours.' Later, at v. 37,
   which describes Jacob's device, the
   marginal note states: 'Jacob herein
   used no deceit, for it was God's commandment', and refers to 31: 9–10
   and a similar gloss.
87 **if . . . not** In Gen. 31: 5–9 Jacob
   further justifies his action to Leah and
   Rachel, claiming that their father
   Laban had cheated him in his wages.
   But Brown, citing Philip Caesar's *Discourse* again, says usury was equated
   with theft.
88–92 **This . . . rams** Antonio recognizes
   heaven's role in Jacob's good fortune,
   but rejects the analogy with usury.
88 **served for** An objection to usury was
   that it involved no labour.

Was this inserted to make interest good?
Or is your gold and silver ewes and rams?

SHYLOCK

I cannot tell; I make it breed as fast.
But note me, signor—

ANTONIO   Mark you this, Bassanio,

The devil can cite Scripture for his purpose.   95
An evil soul producing holy witness
Is like a villain with a smiling cheek,
A goodly apple rotten at the heart.
O, what a goodly outside falsehood hath!

SHYLOCK

Three thousand ducats. 'Tis a good round sum.   100
Three months from twelve; then, let me see, the rate—

ANTONIO

Well, Shylock, shall we be beholden to you?

SHYLOCK

Signor Antonio, many a time and oft
In the Rialto you have rated me
About my moneys and my usances.   105
Still have I borne it with a patient shrug,
For suff'rance is the badge of all our tribe.
You call me misbeliever, cut-throat, dog,

101 then . . . rate—] CAMBRIDGE (*subs.*; Lloyd *conj.*), NCS; ~∧ let me see, the ~. Q, F,
BROWN   102 beholden] POPE; beholding Q, F   108 cut-throat,] HUDSON; ~∧ Q, F

91 **inserted** i.e. in Scripture or into our
discussion
93 **breed** Shylock uses the term meta-
phorically, with sardonic humour that
evidently annoys Antonio. On Shylock's humour, compared with Iago's
and Richard III's, which also derive
from the Vice of the morality plays, see
Overton, 31.
95 **The devil . . . purpose** In Matt. 4: 6
and Luke 4: 10 the devil cites Scripture as he tempts Jesus. The phrase
was proverbial (Tilley D230).
97 **a villain . . . cheek** Cf. *Hamlet* 1.5.109:
'one may smile and smile and be a
villain.'
98 **A goodly . . . heart** proverbial (Dent,
citing Tilley A291.1).
99 **O . . . hath** Evil seeming fair is a
familiar Renaissance phrase.
100–1 **Three . . . rate** Deliberately musing
aloud, Shylock brings the conversation

back to the point.
104 **rated** berated, with a possible play on
'rate' (100)
105, 113, 116 **moneys** 'sums of money'
(*OED* 4), but often indistinguishable
from the singular; also at 137
106 **shrug** Cf. Marlowe, *Jew of Malta*
2.3.23–4: 'I learned in Florence how
to kiss my hand, | Heave up my shoul-
ders when they call me dog' (Malone,
cited by Brown).
107 **suff'rance** forbearance, perhaps with
a play on 'suffering'
badge distinguishing mark. Whether
Shylock should wear an actual badge
to single him out as a Jew, such as a
yellow cap or tall pointed hat (as in
the 1984 RSC production), is arguable.
Jews in Venice and elsewhere often
had to wear such badges in the Renaissance, as they did under Hitler. Cf.
Overton, 13, who rejects distinctive

And spit upon my Jewish gaberdine,
And all for use of that which is mine own.   110
Well then, it now appears you need my help.
Go to, then. You come to me and you say,
'Shylock, we would have moneys': you say so—
You, that did void your rheum upon my beard
And foot me as you spurn a stranger cur   115
Over your threshold, moneys is your suit.
What should I say to you? Should I not say
'Hath a dog money? Is it possible
A cur can lend three thousand ducats?' Or
Shall I bend low and in a bondman's key,   120
With bated breath and whisp'ring humbleness,
Say this: 'Fair sir, you spat on me on Wednesday last;
You spurned me such a day; another time
You called me dog; and for these courtesies
I'll lend you thus much moneys'?   125

ANTONIO

I am as like to call thee so again,
To spit on thee again, to spurn thee, too.
If thou wilt lend this money, lend it not
As to thy friends; for when did friendship take
A breed for barren metal of his friend?   130

113 moneys' . . . so—] This edition; ~, . . . so: Q, F   116 moneys] Q, F; money Q2
119 can] Q; should F   122] Q, F; *two verse lines breaking after* 'this:' VAR. 1793   123
day; . . . time͜] F; ~͜ . . . ~, Q   130 for] Q; of F

garb for Shylock.

109 **gaberdine** loose cloak or mantle,
which became associated in stage
tradition with Jews, although in *Tempest* 2.2.38 Caliban also wears a gaberdine. In Spain, but not in England,
Jews were required to wear cloaks
down to their feet, a fact that may
account for the stage tradition (Kittredge). Other distinctive 'Jewish'
dress, e.g. red or yellow hats, derives
from external sources, not the play
(Overton, 13).

110 **use** (1) employment, (2) interest on
(Brown, citing *Much Ado* 2.1.260–1)
mine own Cf. Matt. 20: 15: 'Is it not
lawful for me to do as I will with mine
own?' (proverbial: Tilley, Dent 099).

114 **rheum** spittle

120 **bondman's key** serf's tone or attitude; cf. 'purchased slave' 4.1.89
(Merchant)

122 **Say . . . last** An alexandrine, or hexameter line, as at 2.7.5, 7, 9; 2.9.27;
3.2.154; etc. See Wright, 146, and
Abbott 499.

126 **like** likely

130 **A breed . . . metal** Antonio emphasizes his antipathy to usury by the paradox of 'barren' metal 'breeding'. The
argument against usury goes back to
Aristotle (*Politics* 1258^b). Warburton
quotes Meres: 'Usury and increase by
gold and silver is unlawful, because
against nature; nature hath made
them *sterile* and *barren*, usury makes
them *procreative*' (Malone). Kittredge
detects a pun on *tokos*, Greek for 'offspring' and 'interest for money'.

But lend it rather to thine enemy
Who, if he break, thou mayst with better face
Exact the penalty.

SHYLOCK   Why, look you, how you storm!
I would be friends with you and have your love,
Forget the shames that you have stained me with,   135
Supply your present wants, and take no doit
Of usance for my moneys; and you'll not hear me.
This is kind I offer.

BASSANIO
This were kindness.

SHYLOCK
This kindness will I show.
Go with me to a notary, seal me there   140
Your single bond, and, in a merry sport,
If you repay me not on such a day,
In such a place, such sum or sums as are
Expressed in the condition, let the forfeit
Be nominated for an equal pound   145
Of your fair flesh to be cut off and taken
In what part of your body pleaseth me.

133 penalty] Q; penalties F   148 body] Q; bodie it F

132 **break** i.e. his day (Brown), or go bankrupt (Riverside); cf. 3.1.108)

133–4 **Why . . . love** At the mention of a penalty, Shylock alters his approach, as Brown notes, citing T. Bell, *Speculation of Usury* (1596), B3ᵛ: 'so soon as the silly poor man maketh mention of usury . . . [the usurer] abateth his sour countenance, and beginneth to smile . . . and calleth him neighbour and friend.'

136–7 **take . . . usance** In requiring no interest, Shylock treats Antonio as a 'brother'. See Deut. 23: 19–20.

136 **doit** small sum; actually, a Dutch coin worth about half a farthing (Fischer)

138 **kind** kindness. Shylock's rhythms are decidedly different from the others' and accentuate his status as an alien (Barton, 172; cf. Wright, 249–50). Here, as at 157–8 below and 3.1.44–7, his clipped phrases emphasize his slightly unidiomatic speech. Brown and Merchant see a pun on (1) benevolent,

generous, (2) natural; NCS sees one in the following lines.

142 **single bond** Much discussed, but I think Blackstone, *Commentaries* (1765), iv. 340 (cited by Furness) offers the best explanation: 'An obligation, or bond, is a deed whereby the obligor obliges himself . . . to pay a certain sum of money to another at a day appointed. If this be all, the bond is called a single one, *simplex obligatio*; but there is generally a condition added. . . . In case this condition is not performed, the bond becomes forfeited.'

145 **condition** covenant, contract (Onions)

146 **nominated for** named as. Olivier paused here, as if searching for a suitably ludicrous condition consistent with 'merry sport' (142).
     equal exact

147 **fair flesh** 'This suggests Shylock's darker, Oriental hue' (Furness).

148 **In . . . me** Cf. 4.1.230, 250–1, where the bond specifies that part nearest Antonio's heart.

ANTONIO

Content, in faith. I'll seal to such a bond,
And say there is much kindness in the Jew.   150

BASSANIO

You shall not seal to such a bond for me.
I'll rather dwell in my necessity.

ANTONIO

Why, fear not, man; I will not forfeit it.
Within these two months—that's a month before
This bond expires—I do expect return   155
Of thrice three times the value of this bond.

SHYLOCK

O father Abram, what these Christians are,
Whose own hard dealings teaches them suspect
The thoughts of others! (*To Bassanio*) Pray you, tell
me this:
If he should break his day, what should I gain   160
By the exaction of the forfeiture?
A pound of man's flesh taken from a man
Is not so estimable, profitable neither,
As flesh of muttons, beeves, or goats. I say,
To buy his favour, I extend this friendship.   165
If he will take it, so; if not, adieu.
And, for my love, I pray you wrong me not.

ANTONIO

Yes, Shylock, I will seal unto this bond.

SHYLOCK

Then meet me forthwith at the notary's;
Give him direction for this merry bond,
And I will go and purse the ducats straight,   170
See to my house—left in the fearful guard
Of an unthrifty knave—and presently

152 **dwell ... necessity** i.e. remain in need
157–9 **O ... others** Shylock continues his jocularity, but compare 4.1.292: there in an aside he remarks with more biting humour on Christians as husbands.
157 **father Abram** Cf. Luke 16: 24, 30 (Shaheen).
158 **dealings teaches** Verbs often take a singular ending with a plural subject (Abbott 333).

165 **To buy his favour** Shylock is quite candid.
167 **wrong me not** do not impute evil motives to me (Riverside)
171 **purse ... straight** i.e. get the money straightaway. But earlier (51–5) Shylock indicated he would have to depend on Tubal for part of the money.
172 **fearful** i.e. causing anxiety
173 **unthrifty** careless

I'll be with you. *Exit*

ANTONIO   Hie thee, gentle Jew.
The Hebrew will turn Christian; he grows kind.   175
BASSANIO
I like not fair terms and a villain's mind.
ANTONIO
Come on; in this there can be no dismay:
My ships come home a month before the day.

Exeunt

2.1 *A flourish of cornetts. Enter the Prince of*
      Morocco, a tawny Moor all in white, and three
      or four followers accordingly, with Portia,
      Nerissa, and their train

MOROCCO
Mislike me not for my complexion,
The shadowed livery of the burnished sun,
To whom I am a neighbour and near bred.
Bring me the fairest creature northward born,
Where Phoebus' fire scarce thaws the icicles,   5
And let us make incision for your love
To prove whose blood is reddest, his or mine.

174 *Exit*] Q, F; *after* 'Jew' CAPELL   175–6 Hie . . . kind] *as in* Q3; *lines break after* 'turne'
Q, F   175 The] Q; This F   176 terms] Q; teames F
2.1] *after* Rowe; *Actus Secundus.* F; *not in* Q   0.1 *A flourish of cornetts.*] MALONE *subs.*;
Flo. Cornets. F (*after* 'traine'); *not in* Q   0.1 *the Prince of*] CAPELL; *not in* Q, F

176 **The Hebrew . . . Christian** A fore-
    shadowing of 4.1.383, perhaps: see
    Lewalski, 334.
    kind (1) generous, (2) natural, as at
    138–40 above; also 'related by kin-
    ship' (Overton, 32).
2.1.0.1 **flourish** fanfare used to announce
    approach (or exit, 46) of a distin-
    guished person
    cornetts in the fifteenth and sixteenth
    centuries, wind instruments made of
    wood (not of brass, like modern cor-
    nets)
  0.2 **tawny** dark, but apparently lighter
    than a blackamoor
    all in white ceremonial colour in
    Islam, here used to contrast with the
    rich colours of Portia's train (NCS)
  0.3 **accordingly** similarly dressed and
    dark skinned
  2 **shadowed livery** dark, distinctive dress

worn by the sun's retainers, i.e. black
skin (Riverside). Morocco's style is replete with metaphors and conceits.
  3 **To . . . bred** As the sun was the primate
    among heavenly bodies, Morocco claims
    proximity and near kinship with the
    best.
  5 **Phoebus' fire** i.e. sunshine. Phoebus
    was the sun god.
  6 **make incision** Perhaps an allusion to
    the practice of lovers stabbing them-
    selves and writing 'languishing letters'
    in their blood, described in Jonson's
    Cynthia's Revels 4.1.200–9 and sug-
    gested in *Lear* 2.1.34–5 (Brown); to
    bloodletting for lovesickness (NS, com-
    paring *L.L.L.* 4.3.93–6); or obliquely
    to the flesh bond (NCS).
  7 **reddest** Red blood signified virility and
    courage, then as now.

I tell thee, lady, this aspect of mine
Hath feared the valiant; by my love, I swear
The best-regarded virgins of our clime   10
Have loved it too: I would not change this hue,
Except to steal your thoughts, my gentle queen.

PORTIA
In terms of choice I am not solely led
By nice direction of a maiden's eyes.
Besides, the lott'ry of my destiny   15
Bars me the right of voluntary choosing.
But if my father had not scanted me,
And hedged me by his wit to yield myself
His wife who wins me by that means I told you,
Yourself, renownèd Prince, then stood as fair   20
As any comer I have looked on yet
For my affection.

MOROCCO   Even for that I thank you.
Therefore, I pray you, lead me to the caskets
To try my fortune. By this scimitar
That slew the Sophy and a Persian prince
That won three fields of Sultan Suleiman,   25
I would o'erstare the sternest eyes that look,

18 wit] Q, F; will HANMER, SISSON   25 prince₍ₐ₎] Q, F; Prince, Q2   26 Suleiman]
OXFORD; Solyman Q, F

9 **feared** frightened
13 **terms** respect
14 **nice** scrupulous, fastidious
   direction (1) guidance, (2) point towards which one turns (NCS). Portia disdains superficial attractions—possibly a hint to Morocco, which completely eludes him, and to the audience.
15 **lott'ry of my destiny** game of chance on which my fate depends (Riverside)
17 **scanted** limited
18 **hedged** confined
   wit (1) wisdom, (2) testament, will (H. Hulme, *Explorations in Shakespeare's Language* (1962), 294)
19 **His** i.e. as his
20 **fair** (1) fair-skinned, (2) well, (3) fairly (Overton, 18–19)
24–31 **By this . . . lady** In this speech and elsewhere, Shakespeare both echoes and burlesques Marlowe's 'mighty line'. See James Shapiro, ' "Which is

The Merchant here, and which The Jew?" Shakespeare and the Economics of Influence', *SStud*, 20 (1988), 273. Cf. Leggatt, 129–30.
25 **Sophy** Emperor of Persia
25–6 **a Persian . . . Suleiman** Morocco claims his scimitar killed, besides the Sophy, a valiant Persian prince who had won three battles against Suleiman. Pooler compares the Turk Brusor's boast in Kyd's *Soliman and Perseda* (1592), 1.3.51–4: 'Against the Sophy in three pitched fields, | Under the conduct of great *Suleiman*, | Have I been chief commander of an host, | And put the flint heart Persians to the sword' (cited by Brown). But, as NCS notes, no Persian won three fields against the victorious Turks, led by Suleiman (1520–66), with whom Morocco was allied.
27 **o'erstare** outstare

Outbrave the heart most daring on the earth,
Pluck the young sucking cubs from the she-bear,
Yea, mock the lion when a roars for prey,   30
To win the lady. But alas the while!
If Hercules and Lichas play at dice
Which is the better man, the greater throw
May turn by fortune from the weaker hand:
So is Alcides beaten by his rage;   35
And so may I, blind fortune leading me,
Miss that which one unworthier may attain,
And die with grieving.
PORTIA   You must take your chance,
And either not attempt to choose at all,
Or swear before you choose, if you choose wrong,   40
Never to speak to lady afterward
In way of marriage. Therefore be advised.
MOROCCO
Nor will not. Come, bring me unto my chance.

28 heart] Q2, F; hart Q   30 a] Q; he Q2, F   31 the‸ lady] Q, F, OXFORD; thee, lady
ROWE 1714, POPE +   35 rage] Q, F; page THEOBALD, POPE; wag NS; rogue SISSON

29 **Pluck . . . she-bear** proverbial example of courage (Tilley, Dent S292). Compare 2 Sam. 17: 8, Prov. 17: 12, etc. (Shaheen).

30 **mock . . . prey** Cf. Ps. 104: 21 (Shaheen).
   a he (contraction for 'ha' = he; Onions)

31 **the lady** See Collation. Morocco here probably no longer speaks to Portia directly but, as often in 2.7, declaims publicly (M. Warren, 'A Note on *The Merchant of Venice* II.i.31', *SQ* 32 (1981), 104–5).

32–8 **If Hercules . . . grieving** In 'Shakespeare's Plutarch', *SQ* 10 (1959), 31–2, E. A. J. Honigmann quotes a passage early in the 'Life of Romulus' in which the keeper of Hercules' temple, having little to do, asked the god to play at dice with him on condition that if the keeper won, Hercules would send him good fortune, and if he lost, the keeper would give him a good supper and a fair gentlewoman to lie with. In the preceding paragraph, Plutarch mentions the she-wolf that gave Romulus and Remus suck. Shakespeare evidently associated the incidents in this passage and, as Honigmann explains, later connected them with two ways in which Hercules could be overcome: by fortune (dice) or emotion (rage). Alcides' (i.e. Hercules') 'rage' (35) balances Morocco's 'grieving' (38), just as the dice balance 'blind fortune' (36). Moreover, the rage connects the theme of the good and bad loser, a corollary to the theme of winning and losing announced at the start by Bassanio, 1.1.140–4. But in Plutarch's story Hercules wins; it is in another story—the poisoned shirt of Nessus—that Hercules exhibits his rage (Ovid, *Metamorphoses* 9.98–238). As earlier in the account of the Sophy and Suleiman, Morocco tends to confuse things; hence, 'rage' (35) requires no emendation (see Collation).

32 **Lichas** For the keeper of the temple Morocco uses the name of the servant who gave Hercules the poisoned shirt of Nessus.

35 **Alcides** Greek name for Hercules; see 3.2.55 n.
   rage Morocco conflates Hercules' fury in the shirt of Nessus (32–8) with his reaction to losing at dice.

PORTIA

First, forward to the temple. After dinner
Your hazard shall be made.

MOROCCO   Good fortune then,   45

To make me blest or cursèd'st among men.

⌈*Flourish of cornetts.*⌉ *Exeunt*

2.2   *Enter Lancelot Gobbo the Clown, alone*

LANCELOT Certainly, my conscience will serve me to run
from this Jew my master. The fiend is at mine elbow
and tempts me, saying to me, 'Gobbo, Lancelot Gobbo,
good Lancelot,' or 'good Gobbo,' or 'good Lancelot
Gobbo, use your legs, take the start, run away.' My   5
conscience says, 'No; take heed, honest Lancelot;
take heed, honest Gobbo,' or, as aforesaid, 'honest
Lancelot Gobbo; do not run; scorn running with thy
heels.' Well, the most courageous fiend bids me pack:
'*Via!*' says the fiend; 'Away!' says the fiend. 'Fore   10
the heavens, rouse up a brave mind,' says the fiend,

46.1] *Cornetts.* F (*printed above* '*Exeunt.*' l. 45); *not in* Q
   2.2] *after* Rowe; *not in* Q, F   0.1 *Lancelot Gobbo*] CAPELL; *not in* Q, F   I LANCELOT]
ROWE; *Clowne* Q, F   3–8 Gobbo] Q2; *Iobbe* Q, F; Job F3–4, ROWE   IO *Via*] ROWE; *fia*
Q, F   IO Fore] 'Fore COLLIER I853, NCS; For Q, F   II heavens,] COLLIER; ~∧ Q, F,
BROWN

43 **Nor will not** i.e. I will not propose
   marriage to anyone else if I choose
   wrong

44 **temple** i.e. church, where Morocco will
   take his oath

45 **hazard** Portia seems to hint here (as
   later to Arragon, 2.9.17, and to Bassanio, 3.2.2) on how to choose (note
   the leaden casket's inscription, 2.7.9),
   although only Bassanio appears astute
   enough to take the hint (Oz, 88).

46 **blest** i.e. most blest; the *-est* of the
   second adjective modifies the first, as
   also at *Measure* 4.6.14 (Abbott 398).

2.2.0 *Lancelot Gobbo* Although the name
   is 'Launcelet Iobbe' throughout the
   soliloquy in Q, the first name is modernized according to standard practice,
   a possible pun on 'little knife' (referring to his attempted witty repartee)
   notwithstanding (compare NCS). Since
   the father's surname is spelled 'Gobbo'
   at his entrance (30 SD) and in sub-

sequent speech headings, that spelling
is adopted throughout. 'Gobbo' is the
Italian word for hunchback, but Shakespeare need not have intended a family
of hunchbacks (Merchant), even if Old
Gobbo was one.

   Clown (1) fool, (2) rustic. Lancelot's
humour exemplifies Will Kemp's, who
doubtless created the role.

1 **serve me** i.e. will say nothing against
it (running away), but on the contrary
exhort me to do it (Schmidt). What
follows is an imagined dialogue in
which the devil and conscience each
seek to influence Lancelot, much in
the manner of the struggle for a man's
soul in a morality play like *Everyman*.

5 **start** sudden journey or flight (Onions)

8–9 **scorn . . . heels** comic redundancy:
'scorn' = (1) disdain, (2) kick aside

9 **courageous** lusty, vigorous (*OED* 3)

10 *Via* Italian for 'away'

10–11 **Fore the heavens** a petty oath =
'fore the heavens!

'and run.' Well, my conscience, hanging about the
neck of my heart, says very wisely to me, 'My honest
friend Lancelot'—being an honest man's son, or
rather an honest woman's son, for indeed my father   15
did something smack—something grow to—he had a
kind of taste—well, my conscience says, 'Lancelot,
budge not.' 'Budge,' says the fiend. 'Budge not,' says
my conscience. 'Conscience,' say I, 'You counsel
well;' 'Fiend,' say I, 'you counsel well.' To be ruled by   20
my conscience, I should stay with the Jew my master
who, God bless the mark, is a kind of devil; and to run
away from the Jew, I should be ruled by the fiend who,
saving your reverence, is the devil himself. Certainly
the Jew is the very devil incarnation; and in my con-   25
science—my conscience is but a kind of hard conscience to offer to counsel me to stay with the Jew. The
fiend gives the more friendly counsel: I will run, fiend.
My heels are at your commandment; I will run.
>   *Enter Old Gobbo, blind, with a basket*
GOBBO Master young man, you, I pray you, which is the   30
   way to Master Jew's?
LANCELOT (*aside*) O heavens, this is my true-begotten
   father who, being more than sand-blind, high-gravel-
   blind, knows me not. I will try confusions with him.

16–17 smack—something grow to—he . . . taste—] This edition; ~, ~ ~ ~ ~; ~ . . . ~;
Q; ~: ~ ~ ~; ~ . . . ~; F   20 Fiend . . . well] Q, F; ~ . . . ill Q2   25 incarnation]
Q, F; incarnall Q2   26 but] Q; *not in* F   29 commandment] Q, F; command Q2   29.1
blind] OXFORD; *not in* Q, F   34 confusions] Q, F; conclusions Q2

12–13 **hanging . . . heart** a comic anatom-
   ical absurdity, signifying a clinging,
   affectionate attitude
14–15 **honest . . . honest** (1) honourable,
   (2) chaste
16 **something** somewhat
   smack savour of, be strongly suggest-
   ive of (NS). Modern punctuation
   clarifies Lancelot's comic difficulty
   with the embarrassing facts. NCS suggests a series of bawdy puns starting
   with 'smack' = 'to kiss noisily' (*OED
   v.* 2); compare E. Partridge *Shakespeare's Bawdy* (1947), 189. Note the
   progression from 'smack' to 'taste'.
   grow to (1) advance to (Onions), (2)
   swell (of the penis)
17 **taste** (1) inclination to (*OED* 7), (2)

   enjoyment of (*OED* 7b)
20 **well** See Collation. Q2's pedantic editor
   missed the joke, as again at line 25.
22 **God . . . mark** An exclamatory phrase,
   perhaps originally used to avert an evil
   omen, but then as an apology when
   uttering something offensive, horrible,
   or profane, as in *Two Gentlemen*
   4.4.18–19 (*OED mark sb.* 18).
24 **saving your reverence** an apology, as
   above (22, n.)
25 **incarnation** Lancelot's malapropism
   for 'incarnate'. Cf. Mistress Quickly's
   blunder, *Henry V* 2.3.29–30 (NS).
32 **true-begotten** another humorous confusion, a further example of Lancelot's
   affected speech for comic purpose
33 **sand-blind, high-gravel-blind** Lancelot

GOBBO Master young gentleman, I pray you, which is   35
the way to Master Jew's?

LANCELOT Turn up on your right hand at the next turning, but at the next turning of all on your left, marry,
at the very next turning, turn of no hand, but turn
down indirectly to the Jew's house.   40

GOBBO By God's sonties, 'twill be a hard way to hit. Can
you tell me whether one Lancelot that dwells with him
dwell with him or no?

LANCELOT Talk you of young Master Lancelot? (*Aside*)
Mark me now; now will I raise the waters.—Talk you   45
of young Master Lancelot?

GOBBO No master, sir, but a poor man's son. His father,
though I say't, is an honest exceeding poor man and,
God be thanked, well to live.

LANCELOT Well, let his father be what a will, we talk of   50
young Master Lancelot.

GOBBO Your worship's friend and Lancelot, sir.

LANCELOT But I pray you, *ergo* old man, *ergo* I beseech
you, talk you of young Master Lancelot?

GOBBO Of Lancelot, an't please your mastership.   55

LANCELOT *Ergo* Master Lancelot. Talk not of Master

---

37 up on] Q; vpon F   41 By] F4; Be Q, F, BROWN, NS, NCS   44–9] Q, F; *as verse,*
beginning with 'Talke', *lines breaking* 'Lancelot?', 'sonne.', 'it)', 'man', 'liue.' Q2   52
sir] Q; *not in* F   54 Lancelot?] Q3; ~. Q, F, BROWN

comically invents degrees of blindness:
'sand- blind' = partly blind; 'gravel-
blind' = midway to 'stone-blind', or
total blindess. As an intensive, 'high'
implies that Old Gobbo is close to
'stone-blind'.

34 **confusions** i.e. conclusions, as in *Hamlet* 3.4.179, where it means experi-
ments. But confusions are precisely
what follow.

37–40 **Turn . . . house** Editors since Theobald have noted the origin of the joke
in Terence, *Adelphi* 4.2. Of course,
both men are standing right in front
of Shylock's door, from which Lancelot
has just emerged (NCS). Stage business
may involve Old Gobbo turning and
turning, or being turned, until he is
somewhat dizzy and certainly confused.

41 **sonties** 'Either a diminutive of an old

form "sont" (i.e. saint) or a corruption
of "sante" (i.e. sanctity)' (Onions).

44 **Master** Trying further confusions,
Lancelot refers to himself by a title
reserved for gentry, not mere servants.

45 **raise the waters** (1) increase confusion
(compare 'deep waters' Ps. 69: 2,
cited by *OED* 6c), (2) start tears

49 **well to live** well to do, rich (as in
Winter's Tale 3.3.117). Gobbo comically contradicts what he has just said.

52 **Your . . . sir** a polite formula for declining the title 'Master'; cf. analogous use in *L.L.L* 5.2.568 (Steevens)

53 *ergo* therefore. Scholars notoriously
overworked this Latin tag; hence,
clowns adopted it, as in *All's Well*
1.3.49 (Brown), where Lavatch uses it
in a comic syllogism, and here, where
Lancelot humorously misuses it.

55 **an't** if it

Lancelot, father, for the young gentleman, according
to Fates and Destinies and such odd sayings—the Sisters Three and such branches of learning—is indeed
deceased; or, as you would say in plain terms, gone to   60
heaven.

GOBBO Marry, God forbid! The boy was the very staff of
my age, my very prop.

LANCELOT (*aside*) Do I look like a cudgel or a hovel-post,
a staff or a prop?—Do you know me, father?   65

GOBBO Alack the day, I know you not, young gentleman. But I pray you, tell me, is my boy, God rest his
soul, alive or dead?

LANCELOT Do you not know me, father?

GOBBO Alack, sir, I am sand-blind; I know you not.   70

LANCELOT Nay, indeed, if you had your eyes, you might
fail of the knowing me: it is a wise father that knows
his own child. Well, old man, I will tell you news of
your son. (*Kneels*) Give me your blessing. Truth will
come to light; murder cannot be hid long. A man's   75
son may, but in the end truth will out.

GOBBO Pray you, sir, stand up; I am sure you are not
Lancelot, my boy.

LANCELOT Pray you, let's have no more fooling about it,
but give me your blessing: I am Lancelot, your boy that   80
was, your son that is, your child that shall be.

74 *Kneels*] COLLIER; *not in* Q, F   76 in the end] Q, F; at the length Q2   79 fooling‸]
Q2, F; ~, Q

57 **father** a common term for elderly men,
as in *Lear* 4.5.72
58 **odd** miscellaneous; compare *Richard
III* 1.3.335: 'And thus I clothe my
naked villainy | With odd old ends
stol'n forth of Holy Writ.'
58–9 **Sisters Three** The three old women,
or Fates, of classical mythology who
spin, measure, and finally cut the
thread of a person's life. Like his predi-
lection for proverbs (74–5), Lancelot's
comic tautologies are part of his
affected 'high style'.
60 **deceased** Lancelot indeed 'raises the
waters' (46) here.
62–3 **boy . . . age** Cf. Tobit 10: 4: 'my
son . . . thou staff of our age' (Bishops'
Bible; Shaheen).
64 **hovel-post** Furness quotes Cotgrave, A

*Dictionary of the French and English Ton-
gues* (1611): 'Escraigne, A little hovel,
made of poles set round with their
ends meeting at the top.' Will Kemp
as Lancelot was far from skinny; see
101–2 n., below.
72–3 **wise . . . child** Lancelot reverses the
proverb 'It is a wise child that knows
its own father' (Tilley C309).
74–5 **Truth . . . long** Kyd juxtaposed the
two proverbs (Tilley T591 and
M1315) in *The Spanish Tragedy*
2.6.58–60 (Brown).
79–80 **fooling . . . blessing** At 1.3.69–71
Shylock alluded to the way Isaac was
fooled into giving Jacob his blessing in
Gen. 27: 19–24. The present scene
enacts a comic variation.
80–1 **your boy . . . shall be** Referring to

133

GOBBO I cannot think you are my son.

LANCELOT I know not what I shall think of that. But I am
Lancelot, the Jew's man, and I am sure Margery your
wife is my mother.   85

GOBBO Her name is Margery, indeed. I'll be sworn, if
thou be Lancelot, thou art mine own flesh and blood.
(Feeling the back of Lancelot's head) Lord worshipped
might he be, what a beard hast thou got! Thou hast
got more hair on thy chin than Dobbin my fill-horse   90
has on his tail.

LANCELOT It should seem, then, that Dobbin's tail grows
backward: I am sure he had more hair of his tail than
I have of my face when I last saw him.

GOBBO Lord, how art thou changed! How dost thou and   95
thy master agree? I have brought him a present. How
'gree you now?

LANCELOT Well, well; but for mine own part, as I have
set up my rest to run away, so I will not rest till I have
run some ground. My master's a very Jew. Give him a   100
present?—give him a halter! I am famished in his service; you may tell every finger I have with my ribs.
Father, I am glad you are come. Give me your present
to one Master Bassanio, who indeed gives rare new
liveries. If I serve not him, I will run as far as God has   105
any ground.

88 *Feeling . . . head*] HALLIWELL *subs., after* 'got!' *l.* 89; *not in* Q, F   94 last] Q2; *lost*
Q, F   97 'gree] ₐgree Q, F; agree Q2   101 present?— . . . halter!] ~, . . . ~, Q, F

the three divisions of time (past, present, future), Lancelot confuses the
order of male development (boy, son,
child) and clumsily paraphrases the
Gloria ('As it was in the beginning, is
now, and ever shall be') (Shaheen).

88 *Feeling . . . head* In stage tradition,
Lancelot kneels with the back of his
head to his father, causing the comic
confusion, which in turn recalls once
more the trick Jacob played on Isaac
(Kittredge; see above, 79–80 n.).

90 **fill-horse** cart-horse. The 'fills', or
'thills', were the shafts of a cart, as in
Troilus 3.2.44.

93 **backward** i.e. from long to short (NCS)

93–4 **of . . . of** on . . . on (Abbott 175)

98–9 **I have . . . rest . . . rest** Lancelot

plays on 'to stake all one has' (a gamb-
ling term from primero) and 'to take
up residence', as in *Lear* 1.1.122–3
(Brown).

100 **very** (an intensifer)

101 **halter** hangman's noose

101–2 **I am . . . ribs** 'Lancelot seizes his
father's hand and brings it into con-
tact with the fingers of his own left
hand which are extended rib-like over
his chest' (NS, suggesting traditional
stage business).

102 **you . . . ribs** Again, Lancelot gets
things backwards, as at 72–3 and 80–1.
tell count

103 **me** an ethical dative, here meaning
'for me' (Abbott 220)

105–6 **as far as . . . ground** i.e. to the ends
of the earth

> *Enter Bassanio with Leonardo and a follower or two*

O rare fortune! here comes the man: to him, father;
for I am a Jew if I serve the Jew any longer.

BASSANIO (*to one of his men*) You may do so, but let it be
so hasted that supper be ready at the farthest by five of   110
the clock. See these letters delivered, put the liveries
to making, and desire Graziano to come anon to my
lodging.   *Exit one of his men*

LANCELOT To him, father.

GOBBO God bless your worship.   115

BASSANIO Gramercy. Wouldst thou aught with me?

GOBBO Here's my son, sir, a poor boy—

LANCELOT Not a poor boy, sir, but the rich Jew's man
that would, sir, as my father shall specify—

GOBBO He hath a great infection, sir, as one would say,   120
to serve—

LANCELOT Indeed, the short and the long is, I serve the
Jew, and have a desire, as my father shall specify—

GOBBO His master and he, saving your worship's reverence, are scarce cater-cousins—   125

LANCELOT To be brief, the very truth is that the Jew, having done me wrong, doth cause me, as my father —
being, I hope, an old man—shall frutify unto you—

GOBBO I have here a dish of doves that I would bestow
upon your worship, and my suit is—   130

LANCELOT In very brief, the suit is impertinent to myself,
as your worship shall know by this honest old man;
and, though I say it, though old man, yet (poor man)
my father.

106.1] OXFORD; *after* 'longer.' *l. 108* Q, F   *Leonardo and*] THEOBALD *subs.*; *not in* Q, F
113.1 *Exit . . . men*] Q2; *not in* Q, F   117 boy—] THEOBALD; ~. Q, F   119 specify—]
THEOBALD 1740; ~. Q, F   121 serve—] JOHNSON; ~. Q, F   123 specify—] THEOBALD
1740; ~. Q, F   125 cater-cousins—] catercosins, Q; catercosins. Q2, F   128 you—]
THEOBALD 1740; ~. Q, F   130 is—] Q2; ~. Q, F   133 (poor man)] BROWN; ₍~ ~₎ Q, F

108 **I am a Jew** proverbial; Jew = 'villain'
(Dent, 145, cites many instances), with
a play on Jew = 'Shylock'

116 **Gramercy** thank you; literally, [God]
grant [you] mercy

118 **poor boy . . . rich Jew's man** Lancelot
delights in such antitheses and in in-
terrupting his father.

120 **infection** Gobbo's malapropism for
'affection', as in *Merry Wives* 2.2.112

(Onions)

125 **scarce** (1) scarcely, (2) stingy (Kittredge, Brown)
cater-cousins close friends

128 **frutify** Lancelot entangles the meanings of *certify*, *notify*, and *fructify*, but
in doing so he prompts his father to
offer his gift.

131 **impertinent** Lancelot of course means
pertinent.

BASSANIO One speak for both. What would you?   135
LANCELOT Serve you, sir.
GOBBO That is the very defect of the matter, sir.
BASSANIO (*to Lancelot*)
    I know thee well. Thou hast obtained thy suit.
    Shylock thy master spoke with me this day
    And hath preferred thee, if it be preferment   140
    To leave a rich Jew's service to become
    The follower of so poor a gentleman.
LANCELOT The old proverb is very well parted between
    my master Shylock and you, sir: you have the grace of
    God, sir, and he hath enough.   145
BASSANIO
    Thou speak'st it well.—Go, father, with thy son.
    Take leave of thy old master and inquire
    My lodging out. (*To one of his men*) Give him a livery
    More guarded than his fellows': see it done.
LANCELOT Father, in. I cannot get a service, no, I have   150
    ne'er a tongue in my head! (*Looking at his palm*) Well,
    if any man in Italy have a fairer table which doth offer
    to swear upon a book, I shall have good fortune. Go to,
    here's a simple line of life; here's a small trifle of

143, 150 LANCELOT] Q2 *subs.*; *Clowne*. Q, F   151 head! Well,] ~. ~, Q2; ~, wel: Q,
F   *Looking . . . palm*] HANMER *subs.*, *after* 'Well'; *not in* Q, F

---

137 **defect** Gobbo's malapropism for *effect* = purport.

140 **preferred** recommended. Shylock gives his reasons, 2.5.45–50.

143 **old proverb** 'The grace of God is gear enough' (Tilley G393), from 2 Cor. 12: 9: 'My grace is sufficient for thee' **parted** shared.

149 **guarded** ornamented with braid or other trim, not necessarily the fool's yellow-guarded coat (Brown, following Leslie Hotson, *Shakespeare's Motley* (1952), 57–62), even if Lancelot in some sense assumes the role of Bassanio's jester.

150–1 **I cannot . . . head** Lancelot speaks sarcastically, as if his father had argued he could not speak for himself.

151–3 **Well, if . . . fortune** As usual, Lancelot gets tangled up in his sentence, but he seems to mean: 'Well, if there is any man in Italy who has a fairer palm [*table*] than I have with

which to swear upon a Bible [*book*] that he will have good fortune—[I don't know who he is].' Or perhaps (the wish fathering the thought) in the logical conclusion of the sentence, 'he shall have good fortune', Lancelot substitutes the first person pronoun (Pooler, cited by Brown). For swearing on a book, see *L.L.L.* 4.3.248–50.

152 **table** In palmistry, the quadrangle formed by the four main lines in the palm of the hand (Onions).

154 **simple** unremarkable, humble. Lancelot speaks ironically here and in the following lines, playing on 'simple' repeatedly.
line of life The 'life' line in the palm is the circular one at the base of the thumb that supposedly indicates the nature or duration of one's life.

154–5 **here's . . . wives** Long and deep lines from the ball of the thumb ('Mount of Venus') towards the line of

wives—alas, fifteen wives is nothing. Eleven widows   155
and nine maids is a simple coming-in for one man;
and then to 'scape drowning thrice, and to be in peril
of my life with the edge of a feather-bed—here are
simple scapes. Well, if Fortune be a woman, she's a
good wench for this gear. Father, come, I'll take my
leave of the Jew in the twinkling.

                              Exit with Old Gobbo

BASSANIO

I pray thee, good Leonardo, think on this:
These things being bought and orderly bestowed,
Return in haste, for I do feast tonight
My best-esteemed acquaintance. Hie thee; go.   165

LEONARDO

My best endeavours shall be done herein.
        He parts from Bassiano. Enter Graziano

GRAZIANO

Where's your master?

LEONARDO   Yonder, sir, he walks.   *Exit*

GRAZIANO Signor Bassiano!

BASSIANO Graziano!

GRAZIANO

I have a suit to you.

BASSIANO   You have obtained it.   170

GRAZIANO

You must not deny me. I must go with you to Belmont.

BASSIANO

Why then, you must. But hear thee, Graziano;

161 twinkling.] Q, F; ~ of an eye Q2   *with Old Gobbo*] ROWE *subs.*; *Exit Clowne* Q, F
166.1 *He . . . Bassiano*] NS *subs.*; *not in* Q, F   167 *Exit*] THEOBALD *subs.*; *after l.* 166 Q,
F (*Exit Leonardo.*)   170 a suit] Q2, F; sute Q

life indicate how many wives a man
will have (Halliwell, quoting Saunder's
Chiromancie; cited by Furness).

156 **coming-in** income, with a sexual
innuendo. Perhaps Lancelot expects
dowries from all these wives (Kittredge).

159 **scapes** escapes. Lancelot evidently alludes to sexual adventures as he 'reads'
his palm.

160 **gear** business

161 **the twinkling** i.e. of an eye, as in 1

Cor. 15: 52 (Shaheen). Q2 unneces-
sarily amplifies.

163 **bestowed** i.e. on the ship he will take
to Belmont

166 **herein** Leonardo refers to the list Bassanio gives him.

170 **a suit** The article, which Q omits and
Q2/F restore, helps the metre, although the short speeches (167–70)
are possibly in prose.
    You . . . it Bassiano's instant agree-
ment seems too offhand to Graziano to
be taken seriously.

Bassanio 2.2

is afraid
his uncooth
friend
will devour
his chance
of winning
Part of ya

The Merchant of Venice

Thou art too wild, too rude and bold of voice—
Parts that become thee happily enough,
And in such eyes as ours appear not faults;   175
But where thou art not known, why, there they show
Something too liberal. Pray thee, take pain
To allay with some cold drops of modesty
Thy skipping spirit, lest through thy wild behaviour
I be misconstered in the place I go to,   180
And lose my hopes.

GRAZIANO   Signor Bassanio, hear me:
If I do not put on a sober habit,
Talk with respect, and swear but now and then,
Wear prayer-books in my pocket, look demurely—
Nay more, while grace is saying hood mine eyes   185
Thus with my hat, and sigh, and say 'Amen',
Use all the observance of civility,
Like one well studied in a sad ostent
To please his grandam—never trust me more.

BASSANIO
Well, we shall see your bearing.   190

GRAZIANO
Nay, but I bar tonight: you shall not gauge me
By what we do tonight.

BASSANIO   No, that were pity.
I would entreat you rather to put on
Your boldest suit of mirth, for we have friends
That purpose merriment. But fare you well.   195
I have some business.

GRAZIANO
And I must to Lorenzo and the rest.
But we will visit you at suppertime.

Exeunt severally

Graziano assures Bassanio that he will be civilized and respectable

176 thou art] Q; they are F   198 *Exeunt severally*] OXFORD; *not in* Q, F

173 **rude** coarse, uncivil (Schmidt)
174 **Parts** qualities
177 **liberal** unrestrained, licentious
178 **modesty** moderation
179 **skipping** flighty, thoughtless
180 **misconstered** misconstrued (a common sixteenth-century variant spelling)
182 **habit** (1) dress, (2) demeanour
184 **Wear . . . pocket** an ostentatious (and often hypocritical) religiosity

185–6 **hood . . . hat** i.e. to give the appearance of concentrated devotion. Men kept their hats on indoors (*Hamlet* 5.2.94), but here Graziano takes off his hat and covers his eyes.
188 **sad ostent** serious or solemn appearance
191 **gauge** measure, estimate
193–4 **put on . . . mirth** Bassanio continues Graziano's clothing metaphor.

2.3   *Enter Jessica and Lancelot the Clown*

JESSICA
I am sorry thou wilt leave my father so.
Our house is hell, and thou, a merry devil,
Didst rob it of some taste of tediousness.
But fare thee well; there is a ducat for thee.
And, Lancelot, soon at supper shalt thou see   5
Lorenzo, who is thy new master's guest.
Give him this letter; do it secretly.
And so farewell. I would not have my father
See me in talk with thee.

LANCELOT Adieu. Tears exhibit my tongue. Most beauti-   10
ful pagan, most sweet Jew! If a Christian do not play
the knave and get thee, I am much deceived. But
adieu. These foolish drops do something drown my
manly spirit. Adieu.   *[Exit]*

JESSICA Farewell, good Lancelot.
Alack, what heinous sin is it in me   15
To be ashamed to be my father's child!
But though I am a daughter to his blood,
I am not to his manners. O Lorenzo,
If thou keep promise, I shall end this strife,
Become a Christian and thy loving wife.   *Exit*   20

2.3] *after* Capell; *not in* Q, F   0.1 *Lancelot*] ROWE; *not in* Q, F   9 in] Q; *not in* F
LANCELOT] Q2; *Clowne.* Q, F   11 do] Q, F; *did* F2   13 something] Q; somewhat F   14
Exit] Q2, F; *not in* Q; *after l.* 15 CAPELL, OXFORD

2.3.3 **taste** degree, extent
10 **exhibit** malapropism for *inhibit*, but
like other blunders of this kind, it conveys another kind of sense: his tears
express the feelings his words should.
11 **pagan** Brown suggests the scurrilous
innuendo, 'prostitute', as at *2 Henry
IV* 2.2.145, referring to Doll Tearsheet.
11–12 **If . . . deceived** The F2 editor evidently thought the reference is to Jessica's begetting by a Christian who
cuckolded Shylock; but as Steevens
argues, Lancelot refers to Lorenzo's
elopement and subsequntly getting

Jessica with child: 'do' thus requires
no emendation. The rest of the sentence is ambiguous somewhat in the
manner of 2.2.152–3: 'playing the
knave' suggests being deceived, whereupon Lancelot imagines himself cuckolded.
13 **drops** tears
19 **manners** character
20 **strife** conflict. As Elizabethans took the
marriage vows literally ('man and wife
are become one flesh'), Jessica as
Lorenzo's wife would be of his 'blood',
not Shylock's.

2.4　*Enter Graziano, Lorenzo, Salarino, and Solanio*

LORENZO

Nay, we will slink away in suppertime,

Disguise us at my lodging, and return

All in an hour.

GRAZIANO

We have not made good preparation.

SALARINO

We have not spoke as yet of torchbearers.　　　　　　5

SOLANIO

'Tis vile, unless it may be quaintly ordered,

And better in my mind not undertook.

LORENZO

'Tis now but four o'clock: we have two hours

To furnish us.

Enter Lancelot with a letter

　　　　　　　　Friend Lancelot, what's the news?

LANCELOT　An it shall please you to break up this, it shall　　10

seem to signify.

LORENZO

I know the hand: in faith, 'tis a fair hand,

And whiter than the paper it writ on

Is the fair hand that writ.　　　　　　*Love is white*

GRAZIANO　　　　　　　Love-news, in faith.　*and pure*

LANCELOT　By your leave, sir.　　　　　　　　　　15

LORENZO　Whither goest thou?

2.4] *after* Capell; *not in* Q, F　0.1 *Salarino*] Q, F, NCS; *Salerio* NS, BROWN, OXFORD + 2–3]
CAPELL; *as one line* Q, F　5 as] F4, NS, OXFORD; us Q, F, SISSON, BROWN, NCS, RIVERSIDE
8 o'clock] a clock Q2; of clock Q, F　9 *Enter Lancelot*] *as in* JOHNSON; *after* 'newes' Q,
F; *after* 'houres' (*l. 8*) Q2　*with a letter*] F; *not in* Q　10 An] THEOBALD; And Q, F; If
Q2　10–11 it shall seem] Q; shall it seeme F

2.4.1–9 Nay . . . furnish us Following the
practice of other Tudor masques,
Lorenzo proposes to his friends that
they slip away from the supper-party
that evening and re-enter disguised.
Masquing was an elaborate affair;
masquers often appeared dressed as
foreign notables, accompanied by a
herald, torchbearers (5), and music for
a spectacular entrance (cf. *Much Ado*
2.1.76–89, *L.L.L.* 5.2.96–126, *Romeo*
1.4.3–10). Graziano and the others
fear that since they have not made
adequate preparation, their plan will

misfire, but Lorenzo urges them on.

6 **quaintly ordered** ingeniously arranged

7 **undertook** Elizabethans often used the
past tense for the participle form, as in
Caesar 1.2.50 (Abbott 343).

9 **furnish** provide

10 **this** i.e. the seal on the letter

11 **seem to signify** i.e. tell you (the news).
As usual, Lancelot enjoys affected
speech.

12–14 **fair hand . . . fair hand** (1) beautiful handwriting, (2) beautiful white
hand. White skin was much admired.

15 **By your leave** pardon me

140

LANCELOT Marry, sir, to bid my old master the Jew to sup
tonight with my new master the Christian.

LORENZO
Hold, here, take this. Tell gentle Jessica
I will not fail her. Speak it privately. Go.   20

Exit Lancelot

Gentlemen,
Will you prepare you for this masque tonight?
I am provided of a torchbearer.

SALARINO
Ay, marry, I'll be gone about it straight.

SOLANIO
And so will I.

LORENZO   Meet me and Graziano   25
At Graziano's lodging some hour hence.

SALARINO
'Tis good we do so.   *Exit with Solanio*

GRAZIANO
Was not that letter from fair Jessica?

LORENZO
I must needs tell thee all. She hath directed
How I shall take her from her father's house,   30
What gold and jewels she is furnished with,
What page's suit she hath in readiness.
If e'er the Jew her father come to heaven,
It will be for his gentle daughter's sake;
And never dare misfortune cross her foot   35
Unless she do it under this excuse:
That she is issue to a faithless Jew.

20–2] CAPELL's *lineation; two lines breaking* 'priuately, | Goe' Q, F   20.1 *Exit Lancelot*]
CAPELL *subs.*; ~ *Clowne* Q, F *after l. 23*   25–6] POPE's *lineation; lines break* 'lodging |
Some' Q, F   27 *with Solanio*] CAPELL *subs.; Exit.* Q, F

19 **this** i.e. a tip
20 **Go** Capell's lineation adds an extra
   syllable to an otherwise regular pentameter line but directs the speech
   rightly to Lancelot. Alternatively, 'Go'
   could begin the next short line, with
   Lancelot's exit separating it from
   Lorenzo's address to his friends (Oxford). Although many editors follow Q
   and retain 'Go' as beginning Lorenzo's
   direction to the others, that construction would seem logical if 'Come' were

   used, or the expletive 'Go to'.
22 **prepare you** i.e. prepare yourselves
23 **a torchbearer** i.e. Jessica in disguise
   (32, 39)
34 **gentle** With a pun on *Gentile*. Referring to Charles Butler's *English Grammar* (1634), Kökeritz, 109, says the
   two words were pronounced alike. Cf.
   2.6.51 for similar word-play.
36 **she** misfortune
37 **she** Jessica
   faithless (1) lacking Christianity, (2)

Come, go with me; (*giving Graziano the letter*) peruse
    this as thou goest.
Fair Jessica shall be my torchbearer.   *Exeunt*

2.5   *Enter Shylock the Jew and Lancelot, his man that*
    was, the Clown

SHYLOCK
Well, thou shalt see—thy eyes shall be thy judge
The difference of old Shylock and Bassanio.
—What, Jessica!—Thou shalt not gormandize
As thou hast done with me.—What, Jessica!—
And sleep and snore, and rend apparel out.   5
—Why, Jessica, I say!

LANCELOT   Why, Jessica!

SHYLOCK
Who bids thee call? I do not bid thee call.

LANCELOT Your worship was wont to tell me I could do
nothing without bidding.

    Enter Jessica

JESSICA Call you? What is your will?   10

SHYLOCK
I am bid forth to supper, Jessica:
There are my keys. But wherefore should I go?
I am not bid for love. They flatter me.
But yet I'll go in hate, to feed upon
The prodigal Christian. Jessica, my girl,   15

38 *giving . . . letter*] OXFORD *subs.*; *not in* Q, F   39 *Exeunt*] ROWE; *Exit.* Q, F
  2.5] *after* Capell; *not in* Q, F   0.1 *Shylock the Jew*] BROWN; *Shylock* ROWE; *Iewe* Q, F
Lancelot] Q2; *not in* Q, F   0.2 *his man . . . Clown*] Q, F; *not in* Q2   was,] NCS (*conj.* NS),
OXFORD; ~ $_\wedge$ Q, F   1 SHYLOCK] Q2; *lewe* Q, F   6 LANCELOT] ROWE; *Clowne* Q, F (*throughout
scene*)   8–9] *as in* Q2; *lines break* 'me, | I' Q, F

untrustworthy (NCS)

2.5.0 *his . . . Clown* NS 105 conjectures
that Shakespeare originally wrote *his
man that was* and afterwards added *the
Clown* in the margin, possibly during
rehearsal, to clarify who was meant.
See Collation. Without the comma,
the SD implies that Lancelot's role
changes, but does it?

3 **What, Jessica** Shylock interrupts his
speech repeatedly to summon Jessica
to him.

  gormandize Cf. Lancelot's version,

2.2.101–2.

5 **rend apparel out** ruin clothes by tear-
ing. Perhaps Shylock alludes to tears
caused by growing fat (3).

11 **bid forth** invited (Malone)

14–15 **to feed . . . Christian** Shylock earlier claimed that because of dietary
restrictions he would not eat with
Christians, 1.3.31–5. The inconsistency is necessary for the elopement
plot more than for the 'malignity' of
Shylock's character, which Steevens
argued was thereby heightened.

14 **feed upon** Psychoanalytically oriented

Look to my house. I am right loath to go: *Foreshadowing*
There is some ill a-brewing towards my rest,
For I did dream of money-bags tonight.

LANCELOT I beseech you, sir, go: my young master doth
  expect your reproach.   *malapropism – should be approach,* 20
SHYLOCK So do I his.   *but also means that eventually Shylock will*
LANCELOT And they have conspired together. I will not   *reproach*
  say you shall see a masque; but if you do, then it was   *his master*
  not for nothing that my nose fell a-bleeding on Black-   *Bassanio.* 25
  Monday last at six o'clock i'th'morning, falling out
  that year on Ash Wednesday was four year in th'
  afternoon.

SHYLOCK

What, are there masques? Hear you me, Jessica: *he tells Jessica*
Lock up my doors, and when you hear the drum   *to lock the*
And the vile squealing of the wry-necked fife,   *house and close* 30
Clamber not you up to the casements then,   *the windows*
Nor thrust your head into the public street   *because of the*
To gaze on Christian fools with varnished faces;   *street party*
But stop my house's ears—I mean my casements:
Let not the sound of shallow fopp'ry enter   35

19–20] *as in* POPE; *lines break* 'Maister | Doth' Q, F; 'go, | My' Q2

critics seize upon this passage and similar ones to argue Shylock's cannibalistic tendency. See e.g. Robert Fliess, *Erogeneity and Libido* (New York, 1957; cited by Holland, 234); Leslie Fiedler, *The Stranger in Shakespeare* (New York, 1972), 109–11; and cf. 3.1.50–1.

18 **dream of money-bags** Dreams supposedly go by opposites; to dream of money was traditionally an ill omen (Furness, Kittredge).
  tonight last night
20 **reproach** Shylock (21) plays upon Lancelot's malapropism for *approach* by taking him literally. He misses the further hint about the elopement ('conspired together', 22), which Lancelot then tries to obscure by his verbal smokescreen (Holland, 239–40).
24–5 **nose . . . Monday** Nosebleeds were an ominous sign, especially on church feast days. But 'Black-Monday' was the day *after* Easter Sunday, so-called be-

cause of the dark mist and hail that fell on that day in 1360 (Brown). Lancelot is obviously parodying Shylock's superstitious dream.
24–7 **Black-Monday . . . afternoon** More nonsense. A Monday could hardly fall on a Wednesday, and the time specifications mock all such prognostications.
28 **What . . . masques** Masquing and other such entertainments run counter to Shylock's sober and strict way of life (35–6).
30 **wry-necked** probably describing the position of the musician's head, not his instrument. Drums and fifes were often used in masques (see 2.4.1–9 n.).
32–3 **Nor . . . to gaze** a possible joke on Jessica's name, derived from 'Iscah' (see 'Sources', above), which Elizabethans glossed as 'she that looketh out' (Lewalski)
33 **varnished faces** faces covered with grotesque painted masks

My sober house. By Jacob's staff I swear
I have no mind of feasting forth tonight.
But I will go. Go you before me, sirrah,
Say I will come.

LANCELOT   I will go before, sir.
(*Aside to Jessica*) Mistress, look out at window, for all
        this: 40
              There will come a Christian by
              Will be worth a Jewës eye.   *Exit*

SHYLOCK
   What says that fool of Hagar's offspring, ha?
JESSICA
   His words were 'Farewell, mistress'; nothing else.
SHYLOCK
   The patch is kind enough, but a huge feeder.   45
   Snail-slow in profit, and he sleeps by day
   More than the wildcat. Drones hive not with me;
   Therefore I part with him, and part with him
   To one that I would have him help to waste
   His borrowed purse. Well, Jessica, go in.   50
   Perhaps I will return immediately.
   Do as I bid you; shut doors after you:
   Fast bind, fast find;
   A proverb never stale in thrifty mind.
                                                       Exit
JESSICA
   Farewell; and if my fortune be not crossed,   55
   I have a father, you a daughter, lost.   *Exit*

42 Jewës] Iewes Q, F; Jewess' POPE; Jewës KEIGHTLEY   Exit] ROWE subs.; not in Q, F
46 and] Q; but F   52–3] as in Q2; one line Q, F

36 **Jacob's staff** Jacob left for Padan-arum
   a poor man with only a staff and returned a rich man: see Gen. 32: 10
   and Heb. 11: 21, and cf. 1.3.68–87.
   above.
42 **Jewës eye** proverbial for something
   very valuable (Dent, 9, 146; Tilley
   J53). The disyllabic pronunciation of
   the possessive fits the jigging verse,
   rendering Pope's emendation unnecessary (see Collation and *TC* 325).
43 **Hagar's offspring** a contemptuous reference to Ishmael, son of Abraham's
   Egyptian concubine Hagar. Mother

   and son were later outcast. See Gen.
   21: 9–21.
45 **patch** fool, dolt, as in *Tempest* 3.2.64
46 **profit** improvement (Brown, citing *As
   You Like It* 1.1.6)
47 **wildcat** a nocturnal animal that sleeps
   by day
53 **Fast . . . find** proverbial for keeping
   things secure and thus finding them
   quickly (Tilley B352). In some productions Jessica mouths the words behind
   Shylock's back, indicating that in Shylock's house they are indeed 'never
   stale'.

2.6   *Enter the masquers, Graziano and Salarino*

GRAZIANO

   This is the penthouse under which Lorenzo

   Desired us to make stand.

SALARINO   His hour is almost past.

GRAZIANO

   And it is marvel he outdwells his hour,

   For lovers ever run before the clock.

SALARINO

   O, ten times faster Venus' pigeons fly

   To seal love's bonds new made than they are wont

   To keep obligèd faith unforfeited.

GRAZIANO

   That ever holds. Who riseth from a feast

   With that keen appetite that he sits down?

   Where is the horse that doth untread again   10

   His tedious measures with the unbated fire

   That he did pace them first? All things that are,

   Are with more spirit chasèd than enjoyed.

   How like a younker or a prodigal

2.6] *after* Capell; *not in* Q, F   0.1 *Salarino*] Q1–2, NCS; *Salino* F; Salerio NS, BROWN, OXFORD +   1–2] *as in* Q, F; *lines break* 'which | Lorenzo' Q2   2 make] Q; ~ a F; *omitted* STEEVENS   6 seal] Q; steale F   14 younker] ROWE, OXFORD; younger Q, F, NS, BROWN +

2.6.0 **Salarino** Most editors emend to Solanio or Salerio, assuming that Salarino, if present here, could not witness
the parting of Bassanio and Antonio
that he describes in 2.8. But Salarino,
like Graziano, does not have to exit at
the same time as Lorenzo and Jessica
(59 n.). Oxford adds the SD *with torchbearers*, and has Lorenzo enter *with a
torch* (19 SD; *TC* 325). Since it is a
night-time scene, the directions may
be justified, although masquers apparently did not themselves carry torches
but were accompanied by bearers who
held them (2.4.1–9 n.). Jessica's entrance with a torch at 57 (cf. 40) may
thus be more appropriate.

1 **penthouse** porch, or shelter with a
sloping roof, as in *Much Ado* 3.3.100

2 **Desired . . . past** The two half-lines
equal an alexandrine and probably account for Q *to make stand* instead of F
to make a stand, although *to stand*
would work as well: see Collation.
NCS and Bevington, however, do not

join the half-lines.

5 **Venus' pigeons** doves that draw the
chariot of Venus, as in *Venus* 1190–4.
The joke concerns a lover's greater
eagerness to consummate a new love
than a marriage contract.

7 **obligèd** plighted, pledged

8 **holds** i.e. holds true

10 **untread** retrace, as in *Venus* 908

11 **tedious measures** 'complicated motions
of the horse in the manage: "measures" = lit. paces in a dance' (NS)

14–19 **How . . . wind** Compare the concerns attributed to Antonio by Solanio
and Salarino, 1.1.8–40. Graziano and
Salarino's dialogue, begun in a light
vein, here modulates to a more sombre
one, suggesting a subtext for Jessica
and Lorenzo's later history. See Introduction, p. 74, and 5.1.1–14 n.

14 **younker** young nobleman. As an allusion to the parable of the Prodigal Son
(Luke 15: 11–32), Q/F *younger* may
not require emendation, but it removes
a feeble redundancy—the result,

idea that a voyage on a boat is like a voyage in love

Foreshadowing Antonio's ships fate

The scarfèd bark puts from her native bay,
Hugged and embracèd by the strumpet wind!
How like the prodigal doth she return,
With over-weathered ribs and raggèd sails,
Lean, rent, and beggared by the strumpet wind!

Love diminished with time.

begging goes well and is great

by the end she will be ragged and weather worn

        Enter Lorenzo

SALARINO
Here comes Lorenzo. More of this hereafter.   20

LORENZO
Sweet friends, your patience for my long abode;
Not I but my affairs have made you wait.
When you shall please to play the thieves for wives,
I'll watch as long for you then. Approach.
Here dwells my father Jew.—Ho! who's within?

I'm stealing a wife today and when you need me to I'll help you do the same.

        Enter Jessica above in boy's clothes

JESSICA
Who are you? Tell me, for more certainty,
Albeit I'll swear that I do know your tongue.
LORENZO Lorenzo, and thy love.
JESSICA

Another reference to a "tongue"

Lorenzo, certain, and my love indeed,
For who love I so much? And now who knows   30
But you, Lorenzo, whether I am yours?
LORENZO
Heaven and thy thoughts are witness that thou art.
JESSICA

Use of word (witness)

Here, catch this casket; it is worth the pains.
I am glad 'tis night, you do not look on me,
For I am much ashamed of my exchange.

lingering reality's of uncertainty about her running away

        17 the] Q; a F   19 wind!] ~? Q, F   24 then] Q, F; therein OXFORD   25 Ho] Q2; Howe Q; Hoa F   25.1 *Enter*] CAPELL; *not in* Q, F   *in boy's clothes*] ROWE; *not in* Q, F

ashamed first of being a Jew's how of running away from that life

probably, of compositorial simplifica-
tion (TC 325: see Collation).

15 **scarfèd bark** boat decorated with flags
   or streamers
16, 19 **strumpet** because wanton, i.e.
   sportive, unreliable; alluding to the
   harlots with whom the Prodigal
   wasted his substance (NS)
18 **over-weathered** damaged by long ex-
   posure to weather
21 **abode** i.e. delay

24 **then** Oxford's emendation, 'therein',
   meaning 'in that affair', 'in that circumstance' (*adv.* 2), helps improve the
   metre (TC 325), but a pause before
   'Approach' does as well.
25 **father** i.e. father-in-law (by anticipa-
   tion)
35 **ashamed . . . exchange** embarrassed by
   changing into boy's clothing, 'with a
   possible hint of misgiving about the
   morality of her robbery and elopement' (NCS)

But love is blind, and lovers cannot see
The pretty follies that themselves commit;
For if they could, Cupid himself would blush
To see me thus transformèd to a boy.

LORENZO

Descend, for you must be my torchbearer. 40

JESSICA

What, must I hold a candle to my shames?
They in themselves, good sooth, are too too light.
Why, 'tis an office of discovery, love,
And I should be obscured.

LORENZO   So are you, sweet,
Even in the lowly garnish of a boy. 45
But come at once,
For the close night doth play the runaway,
And we are stayed for at Bassanio's feast.

JESSICA

I will make fast the doors, and gild myself
With some more ducats, and be with you straight. 50

Exit above

GRAZIANO

Now, by my hood, a gentile and no Jew.

LORENZO

Beshrew me but I love her heartily.
For she is wise, if I can judge of her;
And fair she is, if that mine eyes be true;
And true she is, as she hath proved herself; 55

45–7] as in POPE; *two lines breaking* 'once', 'runaway' Q, F; *three lines breaking* 'boy', 'night', 'run-away' Q2   45 lowly] This edition; louely Q, F   50.1 *Exit above*] THEOBALD *subs.*; *not in* Q, F   51 gentile] Q2, OXFORD; gentle Q, F, NS, BROWN +

36 **love is blind** proverbial (Tilley L506)
41 **hold a candle to** (1) illuminate, (2) proverbial for 'stand by and observe' (Tilley C40; cf. *Romeo* 1.4.38: 'I'll be a candle-holder and look on')
42 **light** (1) evident, (2) immodest
45 **lowly** All editors read 'lovely'; but Q's 'louely', an old but possible spelling of *lowly*, or a compositor's misreading of 'lowly', better fits the context than 'lovely' (NCS, which nevertheless reads 'lovely').
garnish outfit, dress (Onions)

47 **close** secretive
doth . . . runaway steals away quickly
49 **gild myself** lit., cover with gold and hence make myself more attractive and valuable
51 **by my hood** possibly a pun on a common oath and the actual hood, or cape, he wears as part of his costume (Malone)
gentile a pun on *gentle*, as at 2.4.34
52 **beshrew me** a mild oath; lit., evil befall me (cf. modern 'damn me')

And therefore, like herself, wise, fair, and true,
Shall she be placèd in my constant soul.
    Enter Jessica
What, art thou come? On, gentlemen; away!
Our masquing mates by this time for us stay.
                              Exit with Jessica

    Enter Antonio
ANTONIO
    Who's there?   60
GRAZIANO
    Signor Antonio!
ANTONIO
    Fie, fie, Graziano; where are all the rest?
    'Tis nine o'clock: our friends all stay for you.
    No masque tonight: the wind is come about.
    Bassanio presently will go aboard.   65
    I have sent twenty out to seek for you.
GRAZIANO
    I am glad on't: I desire no more delight
    Than to be under sail and gone tonight.   *Exeunt*

2.7 ⌐*Flourish of cornetts.*⌐ *Enter Portia, with the*
      Prince of Morocco and both their trains
PORTIA
    Go, draw aside the curtains and discover
    The several caskets to this noble prince.
        The curtains are drawn and three caskets are revealed
    Now make your choice.

58 gentlemen] Q2, F; gentleman Q, NCS   59.1] This edition; *Exit.* Q, F   61 Antonio!]
~ ? Q, F; ~ . Q2   66 I . . . you] Q, F; *not in* Q2   67 GRAZIANO] Q, F; *not in* Q2
2.7] *after* Capell; *not in* Q, F   0.1 *Flourish of cornetts*] MALONE *following* Capell; *not in*
Q, F   *the Prince of*] CAPELL; *not in* Q, F   2.1] ROWE *subs.* (*after l. 3*); *not in* Q, F

59.1 **Exit with Jessica** In his excitement,
taking Jessica by the hand, Lorenzo
calls everyone to come away. But
Antonio's entrance stops Graziano and
apparently also Salarino, who witnesses Bassanio's parting from the
merchant (2.8.36–49). Q's 'gentleman' (58), defended by NCS, is an
obvious error.
60–8 Antonio's entrance heralds a
change in plan (and possibly an alteration in the text: see NCS 171–2). How

much he knows about the elopement
is unclear; cf. 2.8.7–11, 25–6.
2.7.0 **Flourish of cornetts** See Collation.
F misplaced the SD at 77.1, apparently
inserted in copy near the end of the
scene, after the entrance of Salarino
and Solanio at the beginning of the
next (2.8). As Morocco leaves in a
hurry, a 'state exit' may be inappropriate there (Brown), although a flourish appears at the beginning and end
of his first scene (2.1), and the con-

MOROCCO

This first of gold, who this inscription bears:
'Who chooseth me shall gain what many men desire.'   5
The second, silver, which this promise carries:
'Who chooseth me shall get as much as he deserves.'
This third, dull lead, with warning all as blunt:
'Who chooseth me must give and hazard all he hath.'
[*To Portia*] How shall I know if I do choose the right?   10

PORTIA

The one of them contains my picture, Prince.
If you choose that, then I am yours withal.

MOROCCO

Some god direct my judgement! Let me see;
I will survey th'inscriptions back again.
What says this leaden casket?   15
'Who chooseth me must give and hazard all he hath.'
Must give: for what? For lead? Hazard for lead?
This casket threatens. Men that hazard all
Do it in hope of fair advantages.
A golden mind stoops not to shows of dross.
I'll then nor give nor hazard aught for lead.   20
What says the silver with her virgin hue?
'Who chooseth me shall get as much as he deserves.'
As much as he deserves! Pause there, Morocco,
And weigh thy value with an even hand.   25
If thou beest rated by thy estimation,

4 This] Q; The Q2, F   5 many] Q; *not in* F   10 To Portia] This edition; *not in* Q, F
How . . . right?] *repeated in* F *at top of* sig. P2, col. a   14 th'inscriptions] Q; the
inscriptions F   18 threatens. Men] ROWE; ~_∧ men Q, F

cluding flourish may be 'an ironic ac-
companiment' to Morocco's departure
(NCS). Since the action of this scene
follows 2.1, possibly no flourish is required here, and one at the end is all
that is needed.

1 Go Portia speaks to a member of her
train, possibly Nerissa, although a servant draws the curtain at 2.9.3.
discover disclose, reveal

4 who i.e. which; used to avoid caco-
phony here
8 dull . . . blunt a double pun (Brown):
'dull' = (1) dim, not bright, (2) not
sharp, brought out by 'blunt', which

also means (1) plainspoken, (2) coarse,
unpolished, as in *Lucrece* 1300
12 withal '*possibly* with all' (*TC* 325)
13–60 Some . . . may On the pride and
'self-infatuation' that Morocco's speech
reveals, see Danson, 98–104.
20 shows of dross appearances of worth-
lessness
21 nor . . . nor neither . . . nor
22 virgin hue 'silver is the colour of the
moon, and Diana, the virgin goddess,
is the moon goddess' (Kittredge)
25 with . . . hand impartially
26 estimation probably 'reputation' (as in
Errors 3.1.103) rather than 'valu-
ation' (NCS)

Thou dost deserve enough; and yet 'enough'
May not extend so far as to the lady.
And yet to be afeard of my deserving
Were but a weak disabling of myself.   30
As much as I deserve—why, that's the lady!
I do in birth deserve her, and in fortunes,
In graces, and in qualities of breeding;
But more than these, in love I do deserve.
What if I strayed no farther, but chose here?   35
Let's see once more this saying graved in gold:
'Who chooseth me shall gain what many men desire.'
Why, that's the lady! All the world desires her:
From the four corners of the earth they come
To kiss this shrine, this mortal breathing saint.   40
The Hyrcanian deserts and the vasty wilds
Of wide Arabia are as throughfares now
For princes to come view fair Portia.
The wat'ry kingdom, whose ambitious head
Spits in the face of heaven, is no bar   45
To stop the foreign spirits, but they come
As o'er a brook to see fair Portia.
One of these three contains her heavenly picture.
Is't like that lead contains her? 'Twere damnation
To think so base a thought. It were too gross   50
To rib her cerecloth in the obscure grave.

34 deserve.] Q, F; ~ her *conj.* Capell   44 wat'ry] Q2 (watry); waterie Q, F

30 **disabling** disparagement
34 **deserve** *her* is understood; alternative-ly, 'am worthy' (NCS) may be meant
36 **graved** engraved
40 **shrine . . . saint** Again, using preten-tious language, Morocco mixes up a 'shrine' (reliquary containing a rem-nant of a dead saint) with a live saint, a 'mortal breathing' one.
41 **Hyrcanian deserts** area south of the Caspian Sea, famous for its wildness ('desert' = wilderness)
42 **throughfares** made such by the con-stant traffic of suitors
44 **wat'ry kingdom** i.e. the sea, comple-menting 41–2
   ambitious head i.e. tall waves
46 **foreign spirits** a quibble on 'men of courage' and supernatural beings who,

according to superstition, could not travel easily across water (NS)
49 **like** likely
50 **base** ignoble, with a pun on lead as a base metal
51 **rib** enclose. Corpses were normally wrapped in lead. Brown compares Marlowe's Tamburlaine, ordering his wife Zenocrate to be wrapped in gold (2 *Tamburlaine* 2.4.131). See also M. C. Bradbrook, 'Shakespeare's Recollec-tions of Marlowe', Philip Edwards *et al.* (ed.), *Shakespeare's Styles* (Cambridge, 1980), 191, who notes other Marlo-vian echoes.
   cerecloth waxed cloth in which a corpse was wrapped for burial
   obscure dark; accented on first syllable

Or shall I think in silver she's immured,
Being ten times undervalued to tried gold?
O sinful thought! Never so rich a gem
Was set in worse than gold. They have in England   55
A coin that bears the figure of an angel
Stamped in gold, but that's insculped upon;
But here an angel in a golden bed
Lies all within. Deliver me the key.
Here do I choose, and thrive I as I may.   60

PORTIA

There, take it, Prince; and if my form lie there,
Then I am yours.

Morocco unlocks the golden casket

MOROCCO   O hell! What have we here?
A carrion Death, within whose empty eye
There is a written scroll. I'll read the writing.
(*Reads*) All that glisters is not gold;   65
Often have you heard that told.
Many a man his life hath sold
But my outside to behold.
Gilded tombs do worms infold.
Had you been as wise as bold,   70
Young in limbs, in judgement old,
Your answer had not been inscrolled.
Fare you well; your suit is cold.
Cold indeed, and labour lost.
Then farewell heat, and welcome frost!   75

62 yours.] F; ~? Q; ~! BROWN   62.1] ROWE *subs.*; *not in* Q, F   62–4] *as in* CAPELL;
lines break 'death', 'scroule', 'writing' Q, F   65 *Reads*] RIVERSIDE; *not in* Q, F   69 tombs]
CAPELL (*conj.* Johnson); timber Q, F; wood ROWE   74–5] *as in* Q, F *which add speech prefix*
'Mor.'

53 **ten times ... gold** Gold was then ten
times more valuable than silver.
tried assayed

56 **angel** an English coin with the figure
of the archangel Michael, much
punned upon, as in *Much Ado* 2.3.32

57 **insculped upon** engraved on it

59 **key** rhymes with *may* as elsewhere
with *survey* (Kökeritz, 178; Cercignani, 233)

63 **carrion Death** putrefied death's head,
skull. Cf. 40 n. and the irony here
(NCS).

65 **All ... gold** proverbial (Tilley A146)

69 **tombs** Although Q/F 'timber' fits
the context, it is a plausible misreading of MS 'tombes', which is also better metrically. Malone cites 'gilded
tomb' in Sonnet 101; NS compares
Matt. 23: 27: 'whited [margin:
'Or, painted'] tombs, which appear
beautiful outward, but are within full
of dead men's bones, and of all filthi-
ness.'

72 **inscrolled** i.e. written on the scroll

73 **your ... cold** proverbial (Dent, Tilley
S960.1)

75 **farewell ... frost** Morocco inverts the

Portia, adieu. I have too grieved a heart
To take a tedious leave. Thus losers part.

⌈*Flourish of cornetts.*⌉ *Exit with his train*

PORTIA

A gentle riddance. Draw the curtains, go.
Let all of his complexion choose me so.   *Exeunt*

2.8   *Enter Salarino and Solanio*

SALARINO

Why, man, I saw Bassanio under sail.
With him is Graziano gone along,
And in their ship I am sure Lorenzo is not.

SOLANIO

The villain Jew with outcries raised the Duke,
Who went with him to search Bassanio's ship.

SALARINO

He came too late, the ship was under sail.
But there the Duke was given to understand
That in a gondola were seen together
Lorenzo and his amorous Jessica.
Besides, Antonio certified the Duke
They were not with Bassanio in his ship.   10

SOLANIO

I never heard a passion so confused,
So strange, outrageous, and so variable,
As the dog Jew did utter in the streets:
'My daughter! O my ducats! O my daughter!   15
Fled with a Christian! O my Christian ducats!

77.1 *Flourish of cornetts*] DYCE *subs.*; *not in* Q; *Flo. Cornets.* F *after* 2.8.0.1, *probably misplaced   with his train*] DYCE; *not in* Q, F
   2.8] *after* Capell; *not in* Q, F   0.1 *Salarino*] Q, F, NCS; *Salerio* NS +   3 I am] Q, F; Ime Q2   6 came] Q; comes F   8 gondola] THEOBALD; Gondylo Q; Gondilo F

usual saying (Tilley F769). Since the
Prince, if he loses, must never take a
wife (2.1.40–2), he bids farewell to
love (Kittredge).

77 **tedious** Morocco means 'extended' or
'elaborate', but his malapropism is
ironically appropriate.
   part depart

79 **complexion** (1) disposition, tempera-
ment (*OED* 3), (2) skin colour (*OED* 4).
77.1 See 2.7.0 n.

2.8.0 See Collation and 2.6.0 n.
7–9 **the Duke . . . Jessica** Evidently a
smokescreen to deceive Shylock
further. Jessica was in disguise, and
gondolas were constructed to hide pas-
sengers from view (Brown).
12 **passion** passionate outburst
15–22 **My daughter . . . ducats!** Danson,
182, contrasts Barabas's cries of joy
over his daughter and his gold in
Jew of Malta 2.1.47–54.
16 **Christian ducats** either ducats gained

Justice! The law! My ducats and my daughter!
A sealèd bag, two sealèd bags of ducats,
Of double ducats, stol'n from me by my daughter!
And jewels, two stones, two rich and precious stones,   20
Stol'n by my daughter! Justice! Find the girl!
She hath the stones upon her, and the ducats!'
SALARINO
Why, all the boys in Venice follow him,
Crying, 'His stones, his daughter, and his ducats!'
SOLANIO
Let good Antonio look he keep his day,   25
Or he shall pay for this.
SALARINO   Marry, well remembered. *One ship les*
I reasoned with a Frenchman yesterday, *Sank - myke*
Who told me in the narrow seas that part
The French and English there miscarrièd *Antonios*   30
A vessel of our country, richly fraught. *Foreshadowing!*
I thought upon Antonio when he told me,
And wished in silence that it were not his.
SOLANIO
You were best to tell Antonio what you hear;
Yet do not suddenly, for it may grieve him.
SALARINO
A kinder gentleman treads not the earth.   35
I saw Bassanio and Antonio part.
Bassanio told him he would make some speed
Of his return. He answered, 'Do not so;
Slubber not business for my sake, Bassanio,
But stay the very riping of the time;   40
And for the Jew's bond which he hath of me,

39 Slubber] Q2, F; slumber Q

from Christians or ducats now in
Christian hands (NCS)

19 **double ducats** Italian coins worth
twice as much as a single ducat (Fischer)
20 **two stones** The repetition of 'two
stones' conveys a sexual pun, (1)
jewels, (2) testicles, and Shylock's
emasculation. Compare Solanio's
bawdy pun at 3.1.33.
25-32 **Let . . . his** Although Solanio and

Salarino seem aware of the danger
here, in their scene with Shylock they
foolishly mention Antonio's losses
(3.1.39–40).
27 **reasoned** talked
28 **narrow seas** i.e. the English Channel
39 **Slubber** hurry over, act in a slovenly
manner (Onions)
40 **riping . . . time** A favourite Shakespearian theme; cf. 'Ripeness is all'
(*Lear* 5.2.11).

still eating confident that he will be able to pay back the bond

Let it not enter in your mind of love:

Be merry, and employ your chiefest thoughts
To courtship and such fair ostents of love
As shall conveniently become you there.'   45
And even there, his eye being big with tears,
Turning his face, he put his hand behind him,
And with affection wondrous sensible
He wrung Bassanio's hand; and so they parted.

SOLANIO

I think he only loves the world for him.   *Antonio really [?] Bassanio*   50
I pray thee let us go and find him out
And quicken his embracèd heaviness
With some delight or other.

SALARINO   Do we so.   *Exeunt*

2.9   *Enter Nerissa and a Servitor*

NERISSA

Quick, quick, I pray thee; draw the curtain straight.
similar to / Arrogant The Prince of Arragon hath ta'en his oath,
And comes to his election presently.

*The servitor draws aside the curtain, revealing the
three caskets. ⌈Flourish of cornetts.⌉ Enter the
Prince of Arragon, his train, and Portia*

PORTIA

Behold, there stand the caskets, noble Prince.
If you choose that wherein I am contained,
Straight shall our nuptial rites be solemnized.   5
But if you fail, without more speech, my lord,
You must be gone from hence immediately.

Portia is growing less polite as she likes this suitor less.

ARRAGON

I am enjoined by oath to observe three things:

2.9] *after* Capell; *not in* Q, F   3.1–2 *The . . . caskets*] OXFORD *following* Rowe *subs.; not
in* Q, F   3.2 *Flourish of cornetts*] F *subs., after* 'Portia'   *the Prince of*] CAPELL; *not in* Q,
F   7 you] Q; thou F

42 **mind of love** i.e. mind preoccupied
with love
44 **ostents of love** demonstrations of your
love
45 **conveniently** properly (Onions)
46 **big** swollen
48 **affection wondrous sensible** emotion
amazingly evident
50 **he . . . him** i.e. Bassanio means the

world to him
52 **quicken . . . heaviness** i.e. cheer him
up. Antonio embraces his sorrow as
one might hug grief. Compare 'rash-
embraced despair', 3.2.109.
2.9.0 *Servitor* servant
1, 6 **straight** straightaway
3 **election** choice

First, never to unfold to anyone   10
Which casket 'twas I chose. Next, if I fail
Of the right casket, never in my life
To woo a maid in way of marriage.
Lastly, if I do fail in fortune of my choice,
Immediately to leave you and be gone.   15

PORTIA

To these injunctions everyone doth swear
That comes to hazard for my worthless self.

ARRAGON

And so have I addressed me. Fortune now
To my heart's hope! Gold, silver, and base lead.
'Who chooseth me must give and hazard all he hath.'   20
You shall look fairer ere I give or hazard.
What says the golden chest? Ha, let me see.
'Who chooseth me shall gain what many men desire.'
'What many men desire.' That 'many' may be meant
By the fool multitude that choose by show,
Not learning more than the fond eye doth teach,
Which pries not to th'interior but, like the martlet,
Builds in the weather on the outward wall,
Even in the force and road of casualty.
I will not choose what many men desire,   30
Because I will not jump with common spirits
And rank me with the barbarous multitudes.
Why then, to thee, thou silver treasure-house;
Tell me once more what title thou dost bear.
'Who chooseth me shall get as much as he deserves.'   35

13–15] *as in* Q, F, POPE, OXFORD; *as three lines breaking* 'marriage:', 'lastly,', 'choice,'
CAMBRIDGE, BROWN, NS; *as two lines breaking* 'lastly,', 'choice,' CAPELL, NCS

14 **Lastly** See Collation for alternative lineation. Metrically, 'Lastly' can easily be included as part of 14. Here, as in Q/F, it forms an alexandrine; 'marriage' (13) is trisyllabic.

18–50 **And so . . . desert** As his speech shows, Arragon is 'the epitome of worldly wisdom, self-deceived and incapable of coming to the open truth of the spirit' (Danson, 103). On his descent from the Guise in Marlowe's *The Massacre at Paris*, see Nicholas Brooke, 'Marlowe as Provocative Agent in Shakespeare's Early Plays', *SSur* 14 (1961), 42.

18 **addressed me** prepared myself
Fortune Good luck

24–5 **That 'many' . . . multitude** i.e. the 'many' on the inscription may refer to the foolish masses of people

25 **By** for

26 **fond** foolish

27–9 **martlet . . . casualty** probably the house-martin or swift, which makes its nest on the outer walls of a building (cf. *Macbeth* 1.6.4–8)

29 **force . . . casualty** power and path of accident

31 **jump** be at one, agree

Arragon choses the silver casket and fails

Audience b2w
OPPicielly Roast that
It's the lead
casket

And well said too; for who shall go about
To cozen fortune, and be honourable
Without the stamp of merit? Let none presume
To wear an undeservèd dignity.
O, that estates, degrees, and offices   40
Were not derived corruptly, and that clear honour
Were purchased by the merit of the wearer!
How many then should cover that stand bare!
How many be commanded that command!
How much low peasantry would then be gleaned   45
From the true seed of honour! And how much honour
Picked from the chaff and ruin of the times
To be new varnished! Well, but to my choice.
'Who chooseth me shall get as much as he deserves.'
I will assume desert. Give me a key for this,   50
And instantly unlock my fortunes here.

    He opens the silver casket

PORTIA
Too long a pause for that which you find there.   *he b first*
ARRAGON
What's here? The portrait of a blinking idiot   *in denial about*
Presenting me a schedule. I will read it.
How much unlike art thou to Portia!   *choosly the wrong*   55
How much unlike my hopes and my deservings!   *casket*
'Who chooseth me shall have as much as he deserves.'
Did I deserve no more than a fool's head?
Is that my prize? Are my deserts no better?

43 bare!] ~? Q, F   44 command!] ~? Q, F   46 honour!] ~? Q, F   47 chaff] Q2, F;
chaft Q   48 varnished!] ~? Q2; ~; Q; ~: F   51.1] DELIUS *subs., following* Rowe *(after
l. 52); not in* Q, F   55 Portia!] ~? Q, F

37 **cozen** cheat
38 **stamp** imprint or seal, as on a certificate
40–8 **O . . . varnished** Arragon self-righteously plays on the theme of true merit.
40 **estates . . . offices** 'These terms move from general to particular forms of distinction: estates of the realm (e.g. nobility), ranks within those estates (e.g. earls), and officials (e.g. the Chancellorship)' (NCS).
41 **clear** magnificent, glorious (Schmidt)
43 **cover** Since subordinates uncovered

before superiors in rank, many would then not have to doff their hats before spurious nobility.
45 **gleaned** culled out
46 **seed** (1) germ (of a plant), (2) offspring
47 **ruin** refuse, rubbish, with a suggestion of 'those who have been "ruined", or made destitute, by the times' (Brown, citing *OED* 2*b* and 6)
48 **new varnished** i.e. refurbished and made bright again (after being buried under 'chaff and ruin')
50 **assume** claim, lay claim to (Onions)
54 **schedule** scroll

PORTIA

To offend and judge are distinct offices 60
And of opposèd natures.

ARRAGON   What is here?

(*Reads*)   The fire seven times tried this;
Seven times tried that judgement is
That did never choose amiss.
Some there be that shadows kiss; 65
Such have but a shadow's bliss.
There be fools alive, iwis,
Silvered o'er, and so was this.
Take what wife you will to bed,
I will ever be your head. 70
So be gone: you are sped.

Still more fool I shall appear
By the time I linger here.
With one fool's head I came to woo,
But I go away with two. 75
Sweet, adieu. I'll keep my oath,
Patiently to bear my wroth.

[*Flourish of cornets.*] *Exeunt Arragon and his train*

PORTIA

Thus hath the candle singed the moth.

62 *Reads*] *Hee reads.* Q2 *after l.* 61; *not in* Q, F   72–7] *as in* Q2; Q, F *add speech prefix* 'Arrag.', 'Ar.' *before l.* 72   77.1 *Flourish of cornets*] OXFORD; *not in* Q, F   *Exeunt . . . traine*] CAPELL *subs.; not in* Q, F

60–1 **To . . . natures** 'No man ought to be judge in his own cause' (Tilley M341). Having made his choice, Arragon now presumes to comment on the justice of its result. But Portia may refer to herself, declining to answer Arragon's questions.

62 **this** i.e. the silver casket

63 **Seven . . . judgement** Cf. Ps. 12: 6: 'The words of the Lord are pure words, as [the] silver, tried in a furnace of earth, fined sevenfold' (Noble).

65 **shadows** insubstantial phantoms, but here possibly pictures, alluding to the practice of kissing portraits (Pooler, cited by Brown)

67 **iwis** certainly, assuredly

68 **Silvered o'er** alluding to the silver ornamentation of court officials (NCS) or to medals decorating uniforms (like the Prince's), rather than to the grey hair of aged fools

69 **Take . . . bed** 'Perhaps the poet had forgotten that he who missed Portia was never to marry any woman' (Johnson).

71 **sped** (1) 'done for' (Pooler), (2) hastened on your way

76–8 **oath . . . wroth . . . moth** Phonological variants probably account for these rhymes (Cercignani, 108. 119). spelled *oath, wroath, moath* in Q/F.

77 **wroth** OED cites Q's *wroath* as a variant of *ruth* ('sorrow, grief') as its only reference. More likely, Q's *wroath* was used to suggest the rhyme (76–8 n.) and means 'anger, rage' (OED *wroth sb.* 1), as in *Troilus* 2.3.182 (F; Q *worth*). Arragon is furious.

78 **Thus . . . moth** proverbial (Tilley F394)

O, these deliberate fools! When they do choose,
They have the wisdom by their wit to lose.   80

NERISSA
  The ancient saying is no heresy,
  Hanging and wiving goes by destiny.

PORTIA
  Come, draw the curtain, Nerissa.
        Nerissa draws the curtain
        Enter Messenger

MESSENGER
  Where is my lady?

PORTIA   Here. What would my lord?

MESSENGER
  Madam, there is alighted at your gate
  A young Venetian, one that comes before   85
  To signify th'approaching of his lord,
  From whom he bringeth sensible regreets;
  To wit, besides commends and courteous breath,
  Gifts of rich value. Yet I have not seen   90
  So likely an ambassador of love:
  A day in April never came so sweet
  To show how costly summer was at hand
  As this fore-spurrer comes before his lord.

PORTIA
  No more, I pray thee. I am half afeard
  Thou wilt say anon he is some kin to thee,   95
  Thou spend'st such high-day wit in praising him.
  Come, come, Nerissa; for I long to see
  Quick Cupid's post that comes so mannerly.

NERISSA
  Bassanio, Lord Love, if thy will it be!   *Exeunt*   100

83.1] OXFORD; *not in* Q, F   100 Bassanio, Lord Love,] ROWE; ~ˌ ~, loueˌ Q, F

79 **deliberate** calculating
80 **They . . . lose** 'They are so wise that, when they have used all their wisdom in deliberating, they choose wrong, and lose' (Kittredge).
82 **Hanging . . . destiny** Tilley W232; compare *Twelfth Night* 1.5.18: 'Many a good hanging prevents a bad marriage' (Dent, 245).
84 **my lord** Portia jokingly plays on the messenger's 'my lady'.

88 **sensible regreets** tangible greetings, i.e. gifts (90)
89 **commends** commendations
   breath words
93 **costly** lavish, rich (Onions)
94 **fore-spurrer** forerunner on horseback
97 **high-day wit** i.e. elegant language; *high-day* = 'holiday, festival'
99 **post** messenger
100 See Collation. Although Brown gives a possible defence of Q's pointing (cited

3.1 *Enter Solanio and Salarino*

SOLANIO Now, what news on the Rialto?

SALARINO Why, yet it lives there unchecked that Antonio hath a ship of rich lading wrecked on the narrow seas—the Goodwins I think they call the place—a very dangerous flat and fatal, where the carcasses of many a tall ship lie buried, as they say, if my gossip Report be an honest woman of her word.

SOLANIO I would she were as lying a gossip in that as ever knapped ginger or made her neighbours believe she wept for the death of a third husband. But it is 10 true, without any slips of prolixity or crossing the plain highway of talk, that the good Antonio, the honest Antonio—O that I had a title good enough to keep his name company—

SALARINO Come, the full stop. 15

SOLANIO Ha! What say'st thou? Why, the end is he hath lost a ship.

SALARINO I would it might prove the end of his losses.

SOLANIO Let me say 'amen' betimes, lest the devil cross my prayer, for here he comes in the likeness of a Jew. 20

Enter Shylock

3.1] *after* Rowe; *Actus Tertius* F; *not in* Q   0.1 *Enter*] Q2, F; *not in* Q   *Salarino*] Q, F, NCS; Salerio NS, BROWN, RIVERSIDE, *etc.*   3 wrecked] Q, F (wrackt)   6 gossip] Q; gossips Q2, F   Report] Q3; report Q1–2, F   14 company—] *after* Brown; ~. Q; ~! F   20 my] Q, F; thy THEOBALD   20.1] Q2; *after l.* 22 Q, F

TC 325), Rowe's punctuation is probably right, since Portia has just mentioned Cupid (99).

3.1.2 unchecked not contradicted
3 narrow seas See 2.8.28 n.
Goodwins sandy shoals off the Kentish coast, very dangerous for ships. Several rich cargoes were lost there in 1592–3 (Richard Larn, *Goodwin Sands Shipwrecks* (Newton Abbot, 1977); cited by NCS).
6 tall gallant
gossip 'A person, mostly a woman . . . who delights in idle talk' (*OED* 3). Pooler and others take 'gossip' for a title ('Dame').
9 knapped ginger munched gingersnaps (Merchant)
10 she wept . . . husband The death of a third husband suggests (among other

things) an increased legacy, hence hypocritical mourning.
11 slips of prolixity wordy lies; 'slips' = counterfeit coins (Kittredge)
11–12 crossing . . . talk deviating from straight, honest talk
15 full stop Solanio is not lying, but he is being prolix.
19 betimes quickly, soon
cross thwart, with a quibble on making the sign of the cross at the end of a prayer
20 Enter Shylock Shylock's entrance is marked in many productions not only by a change of disposition—a mixture of grief and anger—but also by a change in dress. In the National Theatre production (1970), for example, he appeared in his shirtsleeves, his formerly elegant attire abandoned.

How now, Shylock! What news among the merchants?

SHYLOCK You knew, none so well, none so well as you, of my daughter's flight.

SALARINO That's certain. I for my part knew the tailor 25
that made the wings she flew withal.

SOLANIO And Shylock for his own part knew the bird
was fledge; and then it is the complexion of them all to
leave the dam.

SHYLOCK She is damned for it.

SALARINO That's certain, if the devil may be her judge.

SHYLOCK My own flesh and blood to rebel!

SOLANIO Out upon it, old carrion, rebels it at these years?

SHYLOCK I say my daughter is my flesh and my blood. 35

SALARINO There is more difference between thy flesh and
hers than between jet and ivory; more between your
bloods than there is between red wine and Rhenish.
But tell us, do you hear whether Antonio have had
any loss at sea or no? 40

SHYLOCK There I have another bad match. A bankrupt, a
prodigal, who dare scarce show his head on the Rialto; a beggar, that was used to come so smug upon the
mart. Let him look to his bond. He was wont to call me

23 knew] Q, F; know Q2   28 fledge] Q (flidge); fledg'd Q2, F   35 my blood] Q; blood
Q2, F

26 **wings** i.e. her page's costume, with a
quibble on 'flight' (24)
28 **fledge** fit to fly
complexion disposition
31 **devil** i.e. Shylock (see 19–20)
32 **flesh and blood** offspring, but Solanio
interprets in a bawdy sense (carnal
appetite)
33 **carrion** i.e. the fleshly nature of man
(*OED* 3b)
38 **red . . . Rhenish** superior white German wine as opposed to common red
wine. Compare 'waterish Burgundy',
Lear 1.1.258 (Merchant).
39–40 **But . . . no** An important turning-point. Why Salarino mentions Antonio's misadventure is unclear, unless
he hopes to taunt Shylock further on
losing his loan. If so, then his hope

backfires. See 2.8.25–32 n.
41 **match** bargain
42 **prodigal** Variously explained. Shylock
may mean Antonio is foolish in lending out money gratis (Thomas Edwards, *Canons of Criticism* (1765);
cited by Furness), or in risking everything in such unsure ventures as the
import-export business (NCS), or in
taking a bond for his friend (Johnson).
43 **smug** smart, spruce, trim (Onions)
mart market, exchange
44 **look . . . bond** The first indication that
Shylock means to collect his pound of
flesh comes precisely here, after Solanio and Salarino have tormented him
about his other loss. He twice repeats
his threat (45–7).

usurer: let him look to his bond. He was wont to lend   45
money for a Christian courtesy: let him look to his
bond.

SALARINO Why, I am sure, if he forfeit, thou wilt not
take his flesh. What's that good for?

SHYLOCK To bait fish withal. If it will feed nothing else, it
will feed my revenge. He hath disgraced me, and
hindered me half a million; laughed at my losses,
mocked at my gains, scorned my nation, thwarted my
bargains, cooled my friends, heated mine enemies. And
what's his reason? I am a Jew. Hath not a Jew eyes?   55
Hath not a Jew hands, organs, dimensions, senses,
affections, passions; fed with the same food, hurt with
the same weapons, subject to the same diseases,
healed by the same means, warmed and cooled by the
same winter and summer, as a Christian is? If you   60
prick us, do we not bleed? If you tickle us, do we not
laugh? If you poison us, do we not die? And if you
wrong us, shall we not revenge? If we are like you in
the rest, we will resemble you in that. If a Jew wrong a
Christian, what is his humility? Revenge. If a Chris-   65
tian wrong a Jew, what should his sufferance be by

---

46 courtesy] Q2, F (curtsie); cursie Q   55 his] Q; the F   65 humility? Revenge.]
BEVINGTON; ~, ~? Q, F

46 **courtesy** act of generosity or benevolence (Brown), with a suggestion of a
small obeisance in return (NS)
50 **bait** i.e. use as bait for
50–1 **If . . . revenge** Cf. 1.3.38–48 and
2.5.14–15 n.
52 **hindered me** i.e. prevented me from
earning
53 **nation** i.e. the Jewish people
cooled alienated
54 **heated** roused up
54–62 **And what's . . . die** These lines,
often quoted out of context, represent
an eloquent apologia, although in context, justifying bloody vengeance, their
appeal is significantly qualified. Shylock, moreover, compares Christian
and Jew 'in terms strictly rational and
reductive: he emphasizes physical properties rather than spiritual and moral
values' (Oz, 92; see also Introduction,
p. 46).
56 **dimensions** bodily parts or proportions,

as in *Lear* 1.2.7 (Onions)
57 **affections, passions** Elizabethan psy-
chology sometimes distinguished between *affections*, originating in sensory
perception, and *passions*, originating in
the emotions.
62 **do . . . die** This is the climax of Shylock's emotional appeal, leading up to
the real point of his speech: revenge.
63 **revenge** Cowden-Clarke recalls Edmund Kean's delivery of this speech:
'the wonderful eyes flashing . . . the
body writhing . . . the arm thrown
upward as witness to the recorded
oath of vengeance. The attitude, as the
voice, rose to a sublime climax when
these words were uttered; then there
was a drop, both of person and tone,
as he hissed out the closing sentence
of deep concentrated malignity' (cited
by Furness).
65 **his humility** i.e. the Christian's kindness, benevolence (sarcastic)

Christian example? Why, revenge. The villainy you
teach me I will execute, and it shall go hard but I will
better the instruction.

 Enter a Man from Antonio

MAN Gentlemen, my master Antonio is at his house and 70
 desires to speak with you both.

SALARINO We have been up and down to seek him.

 Enter Tubal

SOLANIO Here comes another of the tribe. A third cannot
 be matched unless the devil himself turn Jew.

    Exeunt Solanio, Salarino, and Man

SHYLOCK How now, Tubal! What news from Genoa? 75
 Hast thou found my daughter?

TUBAL I often came where I did hear of her, but cannot
 find her.

SHYLOCK Why, there, there, there, there. A diamond
 gone cost me two thousand ducats in Frankfurt. The 80
 curse never fell upon our nation till now; I never felt it
 till now. Two thousand ducats in that and other

---

67 example? Why, revenge.] BEVINGTON; ~, ~‸ ~? Q, F 69.1 *Man*] Q, F; *Servant* ROWE; *Servingman* BROWN, NCS 70 MAN] This edition; Serv. ROWE; *not in* Q, F 72 SALARINO] NCS; *Saleri.* Q; *Salar.* Q2; *Sal.* F 72.1] Q, F; *repeated after l. 74.1* Q1 74.1 *Exeunt . . . Man*] CAPELL (*subs.*); *Exeunt Gentlemen.* Q, F

66 **his sufferance** the Jew's forbearance (cf. 1.3.106)

68 **it . . . hard but** assuredly (Onions); cf. *Hamlet* 3.4.207 (Q2 only): 'and't shall go hard | But I will delve . . .'

72.1 *Enter Tubal* Q repeats this SD after 74, which both Q2 and F correct. Walker attributes the duplication to 'a prompter's having moved back the belated entry at 74 without marking the latter for deletion' (quoted *TC* 325). But this explanation presupposes that copy for Q was a prompt-book (see Textual Introduction, above). Mahood (NCS 172) suggests the error derives from the foul papers: Shakespeare added 69.1–74.1 (to get Salarino and Solanio off) and forgot to delete Tubal's earlier entrance. This is a more plausible explanation than Brown's (the dialogue with Tubal was added later). Note also Q's omission of a speech prefix (70).

74 **be matched** i.e. be found to match them

devil . . . Jew by now a familiar refrain: cf. 2.2.21–2 and 19–20 above

74.1 See Collation. Q's '*Exeunt Gentlemen*' (retained by F), resembles similar SDs in *Hamlet* and is authorial rather than theatrical (Brown).

80 **Frankfurt** site of a famous jewellery fair

80–1 **the curse** See Matt. 27: 25 on Jesus before Pilate and the people of Israel (Noble) and the marginal note in the Geneva Bible: 'If his death be not lawful, let the punishment fall on our heads and our children's, and as they wished, so this curse taketh place to this day.' Compare Luke 13: 34–5 on the destruction of Jerusalem and the marginal note: 'Christ forewarneth them [the Jews] of the destruction of the Temple, and of their whole policy.' Shaheen compares Matt. 23: 37–8 also and Marlowe's *Jew of Malta* 1.2.107–10.

81–2 **I . . . now** The emphasis should be on 'I', as David Suchet spoke it in the 1981 RSC production.

precious, precious jewels. <u>I would my daughter were
dead at my foot</u>, and the jewels in her ear! Would she
were hearsed at my foot and the ducats in her coffin!
No news of them? Why, so. And I know not what's
spent in the search. Why, thou—loss upon loss! The
thief gone with so much, and so much to find the thief,
and no satisfaction, no revenge, nor no ill luck stirring
but what lights o' my shoulders, no sighs but o' my
breathing, no tears but o' my shedding.

TUBAL Yes, other men have ill luck too. Antonio, as I
heard in Genoa—

SHYLOCK What, what, what? Ill luck, ill luck?

TUBAL <u>Hath an argosy cast away, coming from Tripolis.</u>

SHYLOCK I thank God, I thank God. Is it true, is it true?

TUBAL I spoke with some of the sailors that escaped the
wreck.

SHYLOCK I thank thee, good Tubal. Good news, good
news! Ha, ha! Heard in Genoa?

TUBAL Your daughter spent in Genoa, as I heard, one
night fourscore ducats.

SHYLOCK Thou stick'st a dagger in me. I shall never see
my gold again. Fourscore ducats at a sitting! Fourscore ducats!

TUBAL There came divers of Antonio's creditors in my
company to Venice that swear <u>he cannot choose but
break.</u>

86 what's] Q; how much is F   87 thou—] BROWN; ~ˌ Q, F; then F2   90 o' my
shoulders] ROWE 1714; a ~ ~ Q, F; then F2   90 o' my
Q, F; of ~ ~ Q2   91 o' my shedding] ROWE 1714; a ~ ~ Q, F; of ~ ~ Q2   93 Genoa]
ROWE; Genowa? Q, F; Genoway. Q2   94 what?] THEOBALD; ~, Q, F; ~ˌ Q2   luck?]
Q2; ~. Q, F   100 Heard] NEILSON and HILL (*conj.* Kellner); heere Q, F; where? ROWE
101 one] Q, F; in one Q2   104 sitting!] Q2; ~, Q, F   105 ducats!] Q2; ~. Q, F

85 **hearsed** A hearse was 'A light framework of wood used to support the pall
   over the body at funerals' (*OED sb.* 3).
87 **thou—** See Collation. F2 may be right,
   but the Q/F reading with punctuation
   added indicates Shylock's increased
   agitation, dramatically more appropriate than either complacent resignation
   or an apostrophe to loss.
100 **Heard** See Collation. A final *d* could
   easily be misread as *e*, though medial
   ee for *ea* is harder to explain, unless it,

or something like it, stood in the MS
(*TC* 325–6). As emended, Shylock
again repeats Tubal's words (89).
101–2 **Your . . . ducats** Having begun to
   alternate good news and bad, Tubal
   continues to the end of the scene—
   whether deliberately to torment Shylock or not (103, 113) depends upon
   how sadistically the actor represents
   him.
108 **break** go bankrupt

163

SHYLOCK I am very glad of it. I'll plague him, I'll torture
him. I am glad of it.   110

TUBAL One of them showed me a ring that he had of
your daughter for a monkey.

SHYLOCK Out upon her! Thou torturest me, Tubal. It
was my turquoise. I had it of Leah when I was a bachelor. I would not have given it for a wilderness of   115
monkeys.

TUBAL But Antonio is certainly undone.

SHYLOCK Nay, that's true, that's very true. Go, Tubal,
fee me an officer. Bespeak him a fortnight before. I will
have the heart of him if he forfeit, for were he out of   120
Venice I can make what merchandise I will. Go, Tubal,
and meet me at our synagogue. Go, good Tubal; at
our synagogue, Tubal.
                                              Exeunt severally

3.2   *Enter Bassanio, Portia, Graziano, Nerissa, and*
       all their trains. ⸢*The curtains are drawn aside*
       revealing the three caskets⸣

PORTIA (*to Bassanio*)
I pray you, tarry. Pause a day or two
Before you hazard, for in choosing wrong
I lose your company. Therefore forbear awhile.
There's something tells me—but it is not love—

114 turquoise] ROWE; Turkies Q, F   123.1 *Exeunt severally*] OXFORD; *Exeunt.* Q, F
3.2] *after* Rowe; *not in* Q, F   0.1 *Nerissa*] CAPELL; *not in* Q, F   0.2 *trains.*] Q; traine.
F   0.2–3 *The curtain . . . caskets*] OXFORD (NCS subs.); *not in* Q, F

114 **turquoise** 'The turquoise was a natural stone for a betrothal ring . . . for
it was said "to reconcile man and
wife" and faded or brightened with the
wearer's health' (Merchant).
Leah Shylock's wife. In 'The Treatment of Shylock and Thematic Integ-
rity in *The Merchant of Venice*', *SStud* 6
(1970), 78, Albert Wertheim notes
that Jacob's dealings with Laban over
the flocks (see 1.3.68–85) may be motivated partly from revenge, since his
father-in-law tricked him into first taking Leah, not Rachel, as a wife (Gen.
29: 21–30). Leah's only mention occurs here, precisely where Shylock
contemplates revenge against Antonio.
119 **fee . . . officer** i.e. hire a sheriff's officer

or bailiff for me to arrest Antonio
a fortnight before i.e. before Antonio's
due date
121 **merchandise** commercial dealings
122 **synagogue** i.e. presumably where he
will take the oath he refers to at
4.1.225
3.2.1–21 **I pray . . . not I** Quite clearly
in love with Bassanio, Portia finds her
anxiety growing throughout this
speech, leading her to utter silly contradictions (4), revise her request to
delay a day or two (9), confess her
confusion (14–15), and correct herself
embarrassedly (16–20).
2 **in choosing** i.e. if you choose
4 **but . . . love** What else could it be, if
not love, as l. 6 implies?

Portia obviously has far more affection
for Bessanio then any other Suitor

I would not lose you; and you know yourself
Hate counsels not in such a quality.
But lest you should not understand me well—
And yet a maiden hath no tongue but thought—
I would detain you here some month or two
Before you venture for me. I could teach you   10
How to choose right, but then I am forsworn.
So will I never be; so may you miss me.
But if you do, you'll make me wish a sin,
That I had been forsworn. Beshrew your eyes,
They have o'erlook'd me and divided me.   15
One half of me is yours, the other half yours—
Mine own, I would say; but if mine, then yours,
And so all yours. O, these naughty times
Puts bars between the owners and their rights;
And so, though yours, not yours. Prove it so,   20
Let fortune go to hell for it, not I.
I speak too long; but 'tis to piece the time,
To eke it and to draw it out in length
To stay you from election.
BASSANIO   Let me choose,
For as I am, I live upon the rack.   25

Cannot chose him
only desire him

torture devices
Bassanio fells
Torn between his new
love and his old friend.
And cannot
live with
not getting
Portia
stay if
away

16 half yours] Q, F; halfe F2; yours CAPELL   17 if] Q; of F   19 Puts] Q, F; Put F2
20 Prove it so,] OXFORD; (proue ~ ~ ‿) Q, F; ~ ~ not so! CAPELL   22 piece] ROWE
1714 (peece), OXFORD (piece); peize Q, F, KITTREDGE, RIVERSIDE, NCS; peise DYCE, NS,
BROWN, BEVINGTON   23 eke] JOHNSON; ech Q; eck Q2; ich F

6 **quality** manner, style
8 **a maiden . . . thought** i.e. a modest
maiden should only think what she
feels, not speak it; cf. 'as still as any
maid' (Tilley M14.1) and 'maidens
should be seen but not heard' (Tilley
M45)
14 **Beshrew** a mild execration
15 **o'erlook'd** bewitched, as by giving the
evil eye. Cf. *Merry Wives* 5.5.82: 'Vile
worm, thou wast o'erlooked even in
thy birth' (Malone).
17 **would** i.e. should
18 **all yours** Portia at last declares her
true feelings simply.
  naughty wicked, evil, as at 3.3.9,
5.1.91
19 **Puts . . . rights** Portia inveighs against
convention, which puts obstacles between what a person feels and can

declare.
20 **Prove it so** i.e. should it prove so
21 **not I** i.e. I won't be damned for being
forsworn
22 **piece** See Collation. OED records no
spelling 'peize' for *piece* (augment, extend OED 6) or its variant *peise*, but z
for s is often found in 'good' Shakespeare quartos (Brown), and 'piece' fits
better with 'eke' and 'draw it out' (23)
than Q/F 'peize' (or 'peise') = 'hang
weights upon' (a clock and thus make
time go more slowly). Compare *Henry
V* Prol. 23: 'Piece out our imperfec-
tions with your thoughts.'
23 **eke** extend
25 **rack** instrument of torture used, especially in cases of suspected treason
(27, 32–9), to stretch someone out
cruelly until he or she confessed

3.2 The Merchant of Venice

PORTIA
Upon the rack, Bassanio? Then confess
What treason there is mingled with your love.

BASSANIO
None but that ugly treason of mistrust,
Which makes me fear th'enjoying of my love.
There may as well be amity and life   30
'Tween snow and fire as treason and my love.

PORTIA
Ay, but I fear you speak upon the rack,
Where men enforcèd do speak anything.

BASSANIO
Promise me life, and I'll confess the truth.

PORTIA
Well then, confess and live.

BASSANIO   'Confess' and 'love'   35
Had been the very sum of my confession.
O happy torment, when my torturer
Doth teach me answers for deliverance!
But let me to my fortune and the caskets.

[margin: Bassanio can not stand the waiting]

PORTIA
Away, then! I am locked in one of them.   40
If you do love me, you will find me out.
Nerissa and the rest, stand all aloof.
Let music sound while he doth make his choice.
Then if he lose he makes a swanlike end,
Fading in music. That the comparison   45
May stand more proper, my eye shall be the stream
And wat'ry death-bed for him. He may win,

[margin: Portia is clearly trying to increase sense of drama in the scene]

26 Bassanio?] ROWE 1714; ~, Q, F   29 th'enjoying] Q; the enjoying F   33 do] Q; doth F

26–7 **Then . . . love** 'She speaks jestingly, on the general theory that love is never quite faithful or disinterested' (Kittredge).
28 **mistrust** anxiety
29 **fear** be apprehensive of
30–1 **amity . . . fire** compare 'To force fire from snow' (Tilley F284; Dent, 110)
35 **confess and live** Portia first, and then Bassanio, play on the saying 'Confess and be hanged' (Tilley C587; Dent, 78).

42 **aloof** apart, at a distance
43 **Let music sound** This is a departure from the procedure followed in 2.7 and 2.9. Portia's excuses (44–53) lightly camouflage the real reason (see 63–72 n.).
44 **swanlike end** Swans supposedly sang just before their death. Compare *K. John* 5.7.21–2: 'pale faint swan | Who chants a doleful hymn to his own death.'
46 **proper** adapted, suitable (Schmidt)

166

And what is music then? Then music is
Even as the flourish when true subjects bow
To a new-crownèd monarch. Such it is 50
As are those dulcet sounds in break of day
That creep into the dreaming bridegroom's ear
And summon him to marriage. Now he goes,
With no less presence but with much more love
Than young Alcides, when he did redeem 55
The virgin tribute paid by howling Troy
To the sea-monster. I stand for sacrifice;
The rest aloof are the Dardanian wives,
With blearèd visages come forth to view
The issue of th'exploit. Go, Hercules. 60
Live thou, I live. With much much more dismay
I view the fight than thou that mak'st the fray.

[*Here music and*] *a song the whilst Bassanio comments on the caskets to himself*

[ONE FROM PORTIA'S TRAIN]

Tell me where is fancy bred,
Or in the heart or in the head?
How begot, how nourishèd? 65

[ALL]   Reply, reply.

61 live.] JOHNSON; ~ ∧ Q, F   much much] Q; much Q2, F   62 I] Q, F; To Q2   62.1
Here music and] This edition; *Here Musicke* (*as a separate, centred line*) F; *not in* Q   63,
67 ONE . . . TRAIN] OXFORD; *not in* Q, F   66 ALL] NS (*conj.* Lawrence); *not in* Q, F
Reply, reply.] POPE; *printed to right of l. 65* Q, F; *as SD* HANMER, JOHNSON, MALONE; *not
in* ROWE

49 **flourish** fanfare
51 **dulcet sounds** 'An allusion to the custom of playing music under the windows of the bridegroom's bedroom on
the morning of his marriage' (Halliwell, cited by Furness).
54 **presence** noble bearing, dignity
   much more love Alcides fought the sea
   monster not for Hesione's love but for
   the reward of her father's horses (see
   next note).
55 **Alcides** Hercules. In *Metamorphoses*
   11.194–220, Ovid describes how Alcides rescued Hesione from a sea monster sent by the gods as retribution to
   her perfidious father Laomedon, king
   of Troy. Only if this beautiful virgin
   were sacrificed would the floods and
   the ravages of the sea monster stop
   afflicting Troy. In recompense, Her-

cules demanded not the maiden but
Laomedon's famous horses.
58 **Dardanian** Trojan
59 **blearèd** i.e. tear-stained from crying
   (cf. 'howling', 56)
61 **Live thou** i.e. if you live
63 **ONE...TRAIN** The singer is not identified
   in either Q or F but is 'an anonymous
   attendant of no dramatic importance'
   (F. W. Sternfeld, *Music in Shakespearen
   Tragedy* (1963), 105). Richmond Noble
   (*Shakespeare's Use of Song* (1923), cited
   by NS) says the song is a solo and not
   (as in the television version of Jonathan Miller's National Theatre production) a duet.
63–72 **Tell . . . bell** Critics have argued
   over the dramatic function of this
   song. Noting how 'bed', 'head',
   'nourishèd' all rhyme with 'lead',

[ONE FROM PORTIA'S TRAIN] *Appearances are deceiving*

It is engendered in the eyes,
With gazing fed; and fancy dies
In the cradle where it lies.
        Let us all ring fancy's knell: 70
I'll begin it: Ding, dong, bell.
ALL   Ding, dong, bell.

BASSANIO (*addressing the golden casket*)
So may the outward shows be least themselves.
The world is still deceived with ornament.
In law, what plea so tainted and corrupt   *things fair on the*   75
But, being seasoned with a gracious voice,   *outside are not*
Obscures the show of evil? In religion,   *necessarily fair on the*
What damnèd error but some sober brow   *inside*
Will bless it and approve it with a text,
Hiding the grossness with fair ornament?   80
There is no vice so simple but assumes

67 eyes] F; eye Q   68 dies₍₎] ~, F; ~: Q   71 I'll . . . bell] JOHNSON; *as two lines breaking after* 'it.' Q, F   I'll begin it.] *as part of song* JOHNSON; *in roman type, the rest of the song in italic* Q, F   73 *addressing . . . casket*] This edition; *aside* OXFORD; *not in* Q, F   81 vice] F2; voyce Q; voice Q2, F

some see the song as Portia's (or Ne-rissa's) clever hint to Bassanio on how to choose (A. H. Fox-Strangeways, *TLS*, 12 July 1923, p. 472). Others cite the tenor of the verses, their warning against superficial attractions (John Weiss, *Wit and Humour in Shakespeare* (Boston, Mass., 1876), 312). Dover Wilson (NS) notes the tolling bell and the death of fancy as reminders of lead used to 'rib . . . cerecloth in the ob-scure grave' (2.7.51). But others, like Granville-Barker, object to such 'slim tricks'. Brown argues that the lines serve further dramatic purposes: 'The song prevents a third recital of the mottoes on the caskets, dignifies and adds expectation to the dramatic con-text, and prepares the audience for Bassanio's following speech (his thirty-four lines would be an odd elaboration if he believed that the song had given him the secret).' See also Introduction, p. 36, and Peter J. Seng, *The Vocal Songs in the Plays of Shakespeare: A Critical History* (Cambridge, Mass., 1967), 36–43. Seng notes (p. 43) that 'Ding, dong, bell' was a common song burden in Elizabethan times, as in

Ariel's 'Full Fathom Five' in *Tempest*.
63 **fancy** amorous inclination, superficial attraction (opposed to true love), as often in Shakespeare, esp. in *Dream*, where several plots turn on the fantasy (another word for *fancy*) engendered by eyesight. Cf. also *Twelfth Night* 2.4.32.
69 **In the cradle** (1) in the eyes (Capell), or (2) in infancy (Eccles, cited by Fur-ness)
73 **So . . . themselves** external appearan-ces are deceptive
74 **still** ever, continually
76 **seasoned** i.e. rendered more agreeable (Schmidt)
77–9 **In religion . . . text** Cf. Matt. 24: 24: 'For there shall arise false Christs, and false prophets, and shall show great signs and wonders, so that if it were possible, they should deceive the very elect' (Noble). The parallel is not exact; Antonio's warning about the devil's citing Scripture, 1.3.95, is nearer.
81 **vice** Q's 'voyce' (like 'smoyle' for *smile* in Q *Lear* 7.80) is apparently an in-verted spelling on the analogy of words

Some mark of virtue on his outward parts.
How many cowards, whose hearts are all as false
As stairs of sand, wear yet upon their chins
The beards of Hercules and frowning Mars,   85
Who inward searched have livers white as milk?
And these assume but valour's excrement
To render them redoubted. Look on beauty
And you shall see 'tis purchased by the weight,
Which therein works a miracle in nature,   90
Making them lightest that wear most of it.
So are those crispèd, snaky, golden locks
Which makes such wanton gambols with the wind
Upon supposèd fairness, often known
To be the dowry of a second head,   95
The skull that bred them in the sepulchre.
Thus ornament is but the guilèd shore
To a most dangerous sea, the beauteous scarf
Veiling an Indian beauty; in a word,
The seeming truth which cunning times put on   100
To entrap the wisest. Therefore, thou gaudy gold,
Hard food for Midas, I will none of thee.

93 makes] F; maketh Q   99 Indian beauty;] Q, F; ~: ~, COLLIER 1853 (*conj.* Theobald)
101 Therefore,] Q2; ~ then Q, F   102 food] Q, F; foole Q2

like *voyage* spelled 'viage', *Hamlet*
3.3.24 (Cercignani, 247).

82 **his** its
86 **searched** probed, as by a surgeon
    livers . . . milk Elizabethans believed
    the liver, as the seat of courage, ap-
    pears pale when it lacks sufficient
    blood, a sign of cowardice. Compare
    'Milk-livered man', Goneril's accusa-
    tion of cowardice to Albany, *Lear*
    4.2.32.
87 **excrement** i.e. outgrowth of hair, refer-
    ring to 'beards' (85)
88 **redoubted** feared, dreaded
89 **purchased . . . weight** Cosmetics and
    hair (92–6) were bought by the ounce.
91 **lightest** most immodest, quibbling on
    'weight' (89)
92 **crispèd . . . locks** Long, curled, blond
    hair was considered highly attractive,
    as in Botticelli's *Primavera* and his
    portraits of Simonetta.
93 **makes** See Collation. F modernizes and
    thus improves the metre. Singular

verbs often follow plural nouns, especially when they appear as collective
or singular ones (Abbott 333).

95 **dowry** endowment. Cf. Sonnet 68, also
    on the theme of false cosmetic beauty,
    especially 5–7: 'Before the golden
    tresses of the dead, | The right of se-
    pulchres, were shorn away | To live a
    second life on second head.'
97 **guilèd** deceptive, treacherous (from *be-
    guile*)
99 **Indian beauty** ironic; dark skins were
    not favoured by Elizabethans. Compare
    'brow of Egypt', *Dream* 5.1.11, and
    Sonnets 127, 130.
101 **Therefore, thou** See Collation. The
    misprint and correction in Q apparent-
    ly were left side by side (NS). The line
    scans if the first two syllables and the
    fifth and sixth are elided.
102 **Midas** king of Phrygia, whose wish—
    that everything he touched would turn
    to gold—was granted, including his
    food and drink (Ovid, *Metamorphoses*
    11.100–45)

(*To the silver casket*) Nor none of thee, thou pale and
    common drudge
'Tween man and man. But thou, thou meagre lead,
Which rather threaten'st than dost promise aught,   105
Thy paleness moves me more than eloquence,
And here choose I. Joy be the consequence!
PORTIA (*aside*)
How all the other passions fleet to air,
As doubtful thoughts, and rash-embraced despair,
And shudd'ring fear, and green-eyed jealousy!   110
O love, be moderate! Allay thy ecstasy,
In measure rain thy joy, scant this excess!
I feel too much thy blessing. Make it less,
For fear I surfeit.
BASSANIO (*opening the leaden casket*) What find I here?
Fair Portia's counterfeit! What demi-god   115
Hath come so near creation? Move these eyes?
Or whether, riding on the balls of mine,
Seem they in motion? Here are severed lips
Parted with sugar breath. So sweet a bar
Should sunder such sweet friends. Here in her hairs   120
The painter plays the spider and hath woven
A golden mesh t'entrap the hearts of men
Faster than gnats in cobwebs. But her eyes—

103 *To . . . casket*] OXFORD; *not in* Q, F   105 threaten'st] Q; threatnest Q2, F   106
Thy paleness] Q, F; ~ plainness THEOBALD, NS; *Thy* ~ SISSON (*after* Malone)   110
shudd'ring] F; shyddring Q   112 rain] Q, F; range Q2; reine Q3 *corr.*; rein COLLIER
1853 (*conj.* Johnson)   114 *opening . . . casket*] ROWE (*subs.*); *not in* Q, F

103 **common drudge** i.e. because of its use
    in currency
104 **meagre** poor, barren (Schmidt,
    Onions)
106 **paleness** See Collation. Emendation is
    unnecessary, if 'Thy' is stressed (Ma-
    lone). The antithesis to 'eloquence'
    suggests the 'colours' of rhetoric;
    compare *Hamlet* 3.1.55: 'my most
    painted word' (Brown).
109 **As** such as
112 **rain** pour. See Collation. In a meta-
    phor of horsemanship *rein* is possible;
    as homonyms, the two words had in-
    terchangeable spellings. But compare
    Richard II 1.2.6–8: 'the will of heaven
    . . . will rain hot vengeance on the
    offenders' heads.'

    scant limit, restrict (Onions)
115 **counterfeit** portrait
115–16 **What . . . creation** 'The painter
    has come so near to life that he must
    be at least a demigod' (Kittredge). Bas-
    sanio then analyses the features of the
    portrait accordingly.
117 **Or whether** or (Abbott 136)
119–20 **So . . . friends** Alliteration here
    produces an onomatopoeic effect.
122 **golden mesh** a frequent conceit in
    Petrarchan poetry and its imitators;
    cf. 92, above
123–6 **her eyes . . . unfurnished** another
    conceit: a woman's eyes were often
    described as suns or lamps that blinded
    men

How could he see to do them? Having made one,
Methinks it should have power to steal both his   125
And leave itself unfurnished. Yet look how far
The substance of my praise doth wrong this shadow
In underprizing it, so far this shadow
Doth limp behind the substance. Here's the scroll,
The continent and summary of my fortune.   130
(*Reads*)   You that choose not by the view,
          Chance as fair and choose as true!
          Since this fortune falls to you,
          Be content and seek no new.
          If you be well pleased with this,   135
          And hold your fortune for your bliss,
          Turn you where your lady is
          And claim her with a loving kiss.
A gentle scroll. Fair lady, by your leave,
I come by note, to give and to receive.   140
Like one of two contending in a prize
That thinks he hath done well in people's eyes,
Hearing applause and universal shout,
Giddy in spirit, still gazing in a doubt
Whether those peals of praise be his or no,   145
So, thrice-fair lady, stand I, even so,
As doubtful whether what I see be true
Until confirmed, signed, ratified by you.
PORTIA
You see me, Lord Bassanio, where I stand,
Such as I am. Though for myself alone   150
I would not be ambitious in my wish
To wish myself much better, yet for you

126 unfurnished] Q, F; unfinish'd ROWE   131 *Reads*] RIVERSIDE; *not in* Q, F   145 peals]
Q, F; pearles Q2   149 me] Q; my F

125 **it** i.e. the first eye painted
126 **unfurnished** i.e. with a second eye
126–8 **look how far . . . so** see the extent that . . . so
127 **shadow** picture, as at 2.9.65. Bassanio plays on the usual Neoplatonic *substance–shadow* antithesis.
129 **substance** reality, i.e. Portia herself
130 **continent** container
132 **Chance as fair** hazard as fortunately (Riverside)

140 **by note** i.e. authorized by the words in the scroll. Bassanio does not claim his kiss, which still must be 'confirmed, signed, ratified' by Portia (148), though Rowe added an SD for it at the end of 139. The kiss occurs at 167 (NS), or possibly not until after Portia gives him the ring (174), or just before Nerissa breaks in (185), or at several places.
141 **prize** contest, match

171

I would be trebled twenty times myself,
A thousand times more fair, ten thousand times more
    rich,
That only to stand high in your account   155
I might in virtues, beauties, livings, friends
Exceed account. But the full sum of me
Is sum of something which, to term in gross,
Is an unlessoned girl, unschooled, unpractisèd;
Happy in this, she is not yet so old   160
But she may learn; happier than this,
She is not bred so dull but she can learn;
Happiest of all is that her gentle spirit
Commits itself to yours to be directed,
As from her lord, her governor, her king.   165
Myself and what is mine to you and yours
Is now converted. But now I was the lord
Of this fair mansion, master of my servants,
Queen o'er myself; and even now, but now,
This house, these servants, and this same myself   170
Are yours, my lord's. I give them with this ring,
Which when you part from, lose, or give away,
Let it presage the ruin of your love
And be my vantage to exclaim on you.

BASSANIO

Madam, you have bereft me of all words.   175
Only my blood speaks to you in my veins,
And there is such confusion in my powers
As after some oration fairly spoke
By a belovèd prince there doth appear

154 more rich] *as here* COLLIER; *as part of l.* 155 Q, F; *as separate line* MALONE, NS   158
something] Q; nothing F, SISSON   171 lord's] Lords Q; Lord Q2, F

155 **account** estimation
156 **livings** possessions
157 **account** computation. In the follow-
    ing lines Portia continues the language
    of commerce used here and through-
    out the play.
158 **term in gross** cite generally, as a
    whole
165 **lord . . . king** Cf. Katherina's similar
    sentiments at the end of *Shrew* 5.2.
    143, and Eph. 5: 22: 'Wives, submit
    your selves unto your husbands, as

unto the Lord' (NCS). The church
homilies 'Concerning Good Order, and
Obedience to Rulers and Magistrates'
and 'Against Disobedience and Wilful
Rebellion' emphasized this theme (Shaheen).
167, 169 **But** Only, just
174 **vantage** opportunity
    exclaim on denounce, accuse loudly
176 **blood** 'passion' as well as 'vital fluid'
177 **powers** faculties, especially of speech

172

Among the buzzing pleasèd multitude,   180
Where every something being blent together
Turns to a wild of nothing save of joy,
Expressed and not expressed. But when this ring
Parts from this finger, then parts life from hence.
O, then be bold to say Bassanio's dead.   185

NERISSA

My lord and lady, it is now our time
That have stood by and seen our wishes prosper
To cry 'Good joy, good joy, my lord and lady!'

GRAZIANO

My lord Bassanio and my gentle lady,
I wish you all the joy that you can wish,   190
For I am sure you can wish none from me.
And when your honours mean to solemnize
The bargain of your faith, I do beseech you
Even at that time I may be married too.

BASSANIO

With all my heart, so thou canst get a wife.   195

GRAZIANO

I thank your lordship, you have got me one.
My eyes, my lord, can look as swift as yours.
You saw the mistress, I beheld the maid.
You loved, I loved; for intermission
No more pertains to me, my lord, than you.   200
Your fortune stood upon the caskets there,
And so did mine too, as the matter falls;
For wooing here until I sweat again,
And swearing till my very roof was dry
With oaths of love, at last, if promise last,   205
I got a promise of this fair one here

196 have] Q; gaue F   199 loved; for intermission] THEOBALD (*subs.*) ~ ∧ ~ ~, Q, F
201 caskets] Q, F; Casket Q2   204 roof] Q2; rough Q, F; tongue COLLIER 1853

181 **every something** i.e. every bit of utterance
182 **wild** lit. 'wilderness'; here, an unintelligible hubbub
191 **none from me** nothing that I do not wish you (Abbott 158)
193 **bargain** Graziano also uses the language of commerce.
195 **so** provided that

198 **maid** waiting-gentlewoman. Like Maria in *Twelfth Night*, Nerissa is not a menial servant or maid in that sense; hence, she is suitable to marry a gentleman (Merchant).
199 **intermission** delay. Graziano has been as swift as Bassanio.
203 **sweat** (past tense)
204 **roof** i.e. of his mouth

To have her love, provided that your fortune
Achieved her mistress.

PORTIA   Is this true, Nerissa?

NERISSA

Madam, it is, so you stand pleased withal.

BASSANIO

And do you, Graziano, mean good faith?   210

GRAZIANO

Yes, faith, my lord.

BASSANIO

Our feast shall be much honoured in your marriage.

GRAZIANO (*to Nerissa*)

We'll play with them the first boy for a thousand
    ducats.

NERISSA

What, and stake down?

GRAZIANO

No, we shall ne'er win at that sport and stake down.   215
But who comes here? Lorenzo and his infidel?
What, and my old Venetian friend Salerio?

    *Enter Lorenzo, Jessica, and Salerio, a messenger
    from Venice*

BASSANIO

Lorenzo and Salerio, welcome hither;
If that the youth of my new int'rest here
Have power to bid you welcome. (*To Portia*) By your
    leave,   220
I bid my very friends and countrymen,
Sweet Portia, welcome.

PORTIA   So do I, my lord.

209 is, so] Q; ~ so, so F   211 Yes, faith,] ~_∧ ~_∧ Q, F; ~ faith, NCS; ~—~_∧ BROWN
213 *to Nerissa*] OXFORD; *not in* Q, F   217.1] *a . . . Venice*] Q; *not in* F   222-3 So . . .
welcome.] CAPELL; *one line* Q, F

211 **faith** (1) faithfulness, (2) i'faith ('in
    truth')
213 **play** wager
214-15 **stake down** (1) money put down
    in advance (to cover a bet), (2) with
    a limp penis
216 **infidel** Cf. 'stranger', 235 below.
    Jessica has apparently not yet con-
    verted to Christianity, or been fully
    accepted as a Christian. She is not

ignored, as some critics believe (see
    Furness), but greeted generally along
    with the others: later, Graziano pointedly
    asks Nerissa to attend her (235).
217 **Salerio** On his identity, see Textual
    Introduction above, and NCS 179-83.
219-20 **If . . . power** if I, in the first flush
    of my new position here (as lord of the
    manor), have the authority
221 **very** true

They are entirely welcome.

LORENZO

I thank your honour. For my part, my lord,
My purpose was not to have seen you here,   225
But meeting with Salerio by the way
He did entreat me past all saying nay
To come with him along.

SALERIO   I did, my lord,
And I have reason for it. Signor Antonio
Commends him to you.

          He gives Bassanio a letter

BASSANIO   Ere I ope his letter   230
I pray you, tell me how my good friend doth.

SALERIO

Not sick, my lord, unless it be in mind;
Nor well, unless in mind. His letter there
Will show you his estate.

          Bassanio opens the letter and reads to himself

GRAZIANO

Nerissa, cheer yon stranger, bid her welcome.   235
Your hand, Salerio. What's the news from Venice?
How doth that royal merchant, good Antonio?
I know he will be glad of our success;
We are the Jasons, we have won the fleece.

SALERIO

I would you had won the fleece that he hath lost.   240

PORTIA

There are some shrewd contents in yon same paper
That steals the colour from Bassanio's cheek.
Some dear friend dead, else nothing in the world
Could turn so much the constitution
Of any constant man. What, worse and worse?   245

230 *He . . . letter*] THEOBALD (*subs.*); *not in* Q, F   234.1 *Bassanio . . . letter*] ROWE; *open the letter* Q; *He opens the letter* Q2; *Opens the letter* F   *and reads to himself*] This edition; *and reads* OXFORD; *not in* Q, F   235, 241 yon] Q2; yond Q, F

230 **Commends him** sends his greetings
234 **estate** condition
235 **stranger** person one has not seen be-
    fore (*OED* 4); but also alien, or 'non-
    member of society' (*OED* 5; 216 n.;
    3.3.27)

237 **royal** a superlative, like 'princely'
239 **Jasons . . . fleece** See 1.1.169–72 n.
240 **fleece** (1) valuables, (2) fleets
241 **shrewd** evil, cursed
244 **turn** change
245 **constant** self-contained, stable

With leave, Bassanio, I am half yourself,
And I must freely have the half of anything
That this same paper brings you.

BASSANIO   O sweet Portia,
Here are a few of the unpleasant'st words
That ever blotted paper. Gentle lady,   250
When I did first impart my love to you,
I freely told you all the wealth I had
Ran in my veins: I was a gentleman.
And then I told you true. And yet, dear lady,
Rating myself at nothing, you shall see   255
How much I was a braggart. When I told you
My state was nothing, I should then have told you
That I was worse than nothing, for indeed
I have engaged myself to a dear friend,
Engaged my friend to his mere enemy   260
To feed my means. Here is a letter, lady,
The paper as the body of my friend,
And every word in it a gaping wound
Issuing life-blood. But is it true, Salerio,
Hath all his ventures failed? What, not one hit?   265
From Tripolis, from Mexico, and England,
From Lisbon, Barbary, and India,
And not one vessel scape the dreadful touch
Of merchant-marring rocks?

SALERIO   Not one, my lord.
Besides, it should appear that if he had   270
The present money to discharge the Jew,
He would not take it. Never did I know
A creature that did bear the shape of man

---

246 **half yourself** Although the Church insisted on observing the sacraments, betrothal before witnesses (in Elizabethan as in biblical times) was tantamount to marriage. Man and wife being 'one flesh', Portia is half Bassanio.

252–3 **all . . . gentleman** i.e. his lineage, or social position as a gentleman, not material possessions, is all his 'wealth'

255 **Rating** estimating, evaluating

257 **state** estate

259, 260 **engaged** bound, pledged

260 **mere** absolute

262 **as** like, i.e. torn open

265 **Hath** an older form of the third-person plural ending in -*th*; but singular verb forms often occur when they precede plural subjects (Abbott 334, 335)

266 **Mexico** Actually, by Shakespeare's time, Venice's world-wide trade had been curtailed, and it had no direct communication with Mexico (see Furness; McPherson, 52).

268 **scape** escape

269 **merchant-marring** merchant ship destroying

271 **discharge** i.e. pay off the debt to

273–4 **A creature . . . man** i.e. though he

So keen and greedy to confound a man.
He plies the Duke at morning and at night,   275
And doth impeach the freedom of the state
If they deny him justice. Twenty merchants,
The duke himself, and the magnificoes
Of greatest port have all persuaded with him,
But none can drive him from the envious plea   280
Of forfeiture, of justice, and his bond.

JESSICA
When I was with him I have heard him swear
To Tubal and to Chus, his countrymen,
That he would rather have Antonio's flesh
Than twenty times the value of the sum   285
That he did owe him; and I know, my lord,
If law, authority, and power deny not,
It will go hard with poor Antonio.

PORTIA (*to Bassanio*)
Is it your dear friend that is thus in trouble?

BASSANIO
The dearest friend to me, the kindest man,   290
The best-conditioned and unwearied spirit
In doing courtesies, and one in whom
The ancient Roman honour more appears
Than any that draws breath in Italy.

PORTIA
What sum owes he the Jew?   295

BASSANIO
For me three thousand ducats.

PORTIA
                    What, no more?
Pay him six thousand, and deface the bond.

---

283 Chus] Q, F; Cush OXFORD   296–7 What . . . bond] F; *one line* Q

bears the form of a human being, Shylock is more like a fierce ('keen') and ravenous animal in his readiness to destroy a human being

275 **plies** importunes (*OED* 5)

276 **impeach** call in question (Onions)

278 **magnificoes** magnates; cf. 4.1.0 n.

279 **port** social standing
    persuaded pleaded

280 **envious** i.e. motivated by envy, malicious

282–8 **When . . . Antonio** Compare Shylock's soliloquy, 1.3.38–49. Jessica's

jarring words nevertheless go unheeded (NS).

283 **Chus** On this name, see 'Sources', above, p. 22.

291 **best-conditioned** best-natured (Onions)
    unwearied i.e. most unwearied, indefatigable

293 **ancient Roman honour** Shakespeare later dramatized the altruistic spirit of ancient Romans like Brutus, who acted only from the highest motives; see *Caesar* 5.5.67–74.

297 **deface** cancel, obliterate

Double six thousand, and then treble that,
Before a friend of this description
Shall lose a hair through Bassanio's fault.　　　　　　300
First go with me to church and call me wife,
And then away to Venice to your friend;
For never shall you lie by Portia's side
With an unquiet soul. You shall have gold
To pay the petty debt twenty times over.　　　　　　305
When it is paid, bring your true friend along.
My maid Nerissa and myself meantime
Will live as maids and widows. Come, away,
For you shall hence upon your wedding-day.
Bid your friends welcome, show a merry cheer:　　　310
Since you are dear bought, I will love you dear.
But let me hear the letter of your friend.

BASSANIO (*reads*) 'Sweet Bassanio, my ships have all miscarried, my creditors grow cruel, my estate is very
low, my bond to the Jew is forfeit; and since in paying　315
it, it is impossible I should live, all debts are cleared
between you and I, if I might but see you at my death.
Notwithstanding, use your pleasure. If your love do
not persuade you to come, let not my letter.'

PORTIA
O love, dispatch all business, and be gone!　　　　　320

BASSANIO
Since I have your good leave to go away,
I will make haste, but till I come again,
No bed shall e'er be guilty of my stay,
Nor rest be interposer 'twixt us twain.　　　　*Exeunt*

313 BASSANIO (*reads*)] ROWE; *not in* Q, F　　317 but] Q; *not in* F　　320 PORTIA] Q, F; *not in* Q2　　324 Nor] Q, F; No Q2

300 **through** bisyllabic
310 **cheer** countenance, aspect
311 **Since . . . dear** i.e. since Antonio has risked so much for you, I will match him in love. Some critics believe Portia indelicately implies Bassanio is costing her a lot of money (Pope relegated these lines to the foot of the page as 'unworthy of Shakespeare'; Brown, p. lvii). That interpretation is possible; but given the commercial parlance throughout the play, the figurative sense seems uppermost. H. T. Price

(*JEGP* 55 (1956), 644) suggests an allusion to the proverb 'Dear bought and far-fetched are dainties for ladies' (Tilley D12; cited by Brown).
314 **estate** condition
317 **between you and I** This construction seems to have been an Elizabethan idiom (and sadly is becoming a modern one). The sound of *d* and *t* before *me* was avoided (Abbott 205).
318 **use your pleasure** follow your own inclination
323–4 No . . . twain 'Bassanio takes an

3.3 *Enter Shylock the Jew, Solanio, Antonio, and the*
Gaoler

SHYLOCK

Gaoler, look to him. Tell not me of mercy.

This is the fool that lent out money gratis.

Gaoler, look to him.

ANTONIO   Hear me yet, good Shylock.

SHYLOCK

I'll have my bond. Speak not against my bond.

I have sworn an oath that I will have my bond.   5

Thou called'st me dog before thou hadst a cause,

But since I am a dog, beware my fangs.

The Duke shall grant me justice. I do wonder,

Thou naughty gaoler, that thou art so fond

To come abroad with him at his request.   10

ANTONIO

I pray thee, hear me speak.

SHYLOCK

I'll have my bond; I will not hear thee speak.

I'll have my bond; and therefore speak no more.

I'll not be made a soft and dull-eyed fool

To shake the head, relent, and sigh, and yield   15

To Christian intercessors. Follow not.

I'll have no speaking; I will have my bond.   *Exit*

3.3] *after* Rowe; *not in* Q, F   0.1 Shylock] ROWE; *not in* Q, F   Solanio] *and* ~ F; *and*
Salerio Q; *and Salarino* Q2   Antonio] *and* ~ Q, F   1, 4, 12 SHYLOCK] ROWE; *Iew.* Q, F
2 lent] Q; *lends* F   17 Exit] *Exit Iew.* Q, F

oath which is common in old roman-
ces—not to sleep until he has accomplished his undertaking' (Kittredge).

3.3.0 *Shylock the Jew* See Collation. 'The
use of "Jew" in the stage direction[s]
and speech headings of this scene is
significant; Shylock appears only as
the money-lender, not the father'
(NCS). But see Textual Introduction.
Solanio Since Salerio is in Belmont
(3.2), Q's '*Salerio*' is obviously wrong,
probably the result of misunderstand-
ing an abbreviation '*Sal.*' or, more likely,
'*Sol.*' (easily confused in manuscript:
compare Q's '*Sol.*' 18 SH, but '*Sal.*' 24
SH). Both Q2 and F recognize the error

(see Collation). Since F was probably
collated against a theatrical prompt-
book, its reading is preferred.
4–5 **I'll . . . bond** Shylock's attitude has
hardened; thus he harps on his oath
and his bond, here and later (12–17).
6 **dog** See 1.3.108.
8 **The Duke . . . justice** Shylock's confidence is based on another aspect of the
myth of Venice, i.e. its renown for
justice (McPherson, 36–8), as Antonio
recognizes (26–31).
9 **naughty** wicked
fond foolish
10 **abroad** i.e. outside the gaol
14 **dull-eyed** unseeing, i.e. unperceptive,
easily deceived

SOLANIO

  It is the most impenetrable cur

  That ever kept with men.

ANTONIO   Let him alone.

  I'll follow him no more with bootless prayers.   20

  He seeks my life. His reason well I know:

  I oft delivered from his forfeitures

  Many that have at times made moan to me;

  Therefore he hates me.

SOLANIO   I am sure the Duke

  Will never grant this forfeiture to hold.   25

ANTONIO

  The Duke cannot deny the course of law,

  For the commodity that strangers have

  With us in Venice, if it be denied,

  Will much impeach the justice of the state,

  Since that the trade and profit of the city   30

  Consisteth of all nations. Therefore, go.

  These griefs and losses have so bated me,

  That I shall hardly spare a pound of flesh

  Tomorrow to my bloody creditor.

  Well, gaoler, on. Pray God Bassanio come   35

  To see me pay his debt, and then I care not!

                               Exeunt

3.4 *Enter Portia, Nerissa, Lorenzo, Jessica, and*
       Balthasar, a man of Portia's

LORENZO

  Madam, although I speak it in your presence,

  You have a noble and a true conceit

  Of godlike amity, which appears most strongly

  In bearing thus the absence of your lord.

24–5 I . . . hold] POPE; *lines break after* 'grant' Q, F
    3.4] *after* Rowe; *not in* Q, F   0.2 *Balthasar*] THEOBALD; *not in* Q, F

19 **kept** associated
20 **bootless** unavailing
22 **forfeitures** penalties
27 **commodity** commercial privileges
28 **it** i.e. course of law (26)
29 **impeach** Antonio knows that Venice depends upon international trade, which in turn depends upon the city's reputation for justice under law.

32 **bated** diminished, weakened
3.4.2 **conceit** conception, idea
3 **godlike amity** i.e. the highest form of friendship, Platonic love, such as exists between Antonio and Bassanio. Lorenzo and Portia have apparently been talking about them (NS). The terms *love, lover* (7, 13, 17) refer to this concept; cf. NCS 22–4.

But if you knew to whom you show this honour,   5
How true a gentleman you send relief,
How dear a lover of my lord your husband,
I know you would be prouder of the work
Than customary bounty can enforce you.
PORTIA
I never did repent for doing good,
Nor shall not now; for in companions   10
That do converse and waste the time together,
Whose souls do bear an equal yoke of love,
There must be needs a like proportion
Of lineaments, of manners, and of spirit,   15
Which makes me think that this Antonio,
Being the bosom lover of my lord,
Must needs be like my lord. If it be so,
How little is the cost I have bestowed
In purchasing the semblance of my soul   20
From out the state of hellish cruelty.
This comes too near the praising of myself,
Therefore no more of it. Hear other things:
Lorenzo, I commit into your hands
The husbandry and manage of my house   25
Until my lord's return. For mine own part,
I have toward heaven breathed a secret vow
To live in prayer and contemplation,
Only attended by Nerissa here,
Until her husband and my lord's return.
There is a monastery two miles off,   30
And there we will abide. I do desire you
Not to deny this imposition,

20 soul‿] ~; Q, F; ~, Q2   21 cruelty] Q, F; misery Q2   23 Hear] THEOBALD subst.
(Thirlby); heere Q, F; Here are ROWE

9 **Than . . . you** i.e. than your usual acts of benevolence make you perform
12 **waste** spend
13 **bear . . . love** i.e. love mutually
14 **needs** necessarily
15 **lineaments** characteristics. Although the *OED* does not record the figurative sense before 1638, Shakespeare probably intended it here, not physical features.

17 **bosom lover** i.e. close friend
20 **purchasing** redeeming
semblance of my soul i.e. Antonio, the image of her soul (through Bassanio's; see 3.2.246 n.)
25 **husbandry and manage** care and management
30 **her . . . lord's** The possessive pertains to both nouns (Abbott 397).
33 **imposition** charge

The which my love and some necessity
Now lays upon you.
LORENZO   Madam, with all my heart,   35
I shall obey you in all fair commands.
PORTIA
My people do already know my mind
And will acknowledge you and Jessica
In place of Lord Bassanio and myself.
So fare you well till we shall meet again.   40
LORENZO
Fair thoughts and happy hours attend on you!
JESSICA
I wish your ladyship all heart's content.
PORTIA
I thank you for your wish, and am well pleased
To wish it back on you. Fare you well, Jessica.
                    Exeunt Jessica and Lorenzo

Now, Balthasar,   45
As I have ever found thee honest-true,
So let me find thee still. Take this same letter,
And use thou all th'endeavour of a man
In speed to Padua. See thou render this
Into my cousin's hands, Doctor Bellario,   50
And look what notes and garments he doth give thee,
Bring them, I pray thee, with imagined speed
Unto the traject, to the common ferry
Which trades to Venice. Waste no time in words,
But get thee gone. I shall be there before thee.   55

---

40 So fare you well] Q, F; And so farewell Q2   44 Fare you well] Q, F; farewell Q2
44.1] ROWE (*subs.*); *Exeunt.* Q, F   45–6 Now . . . honest-true] POPE; *one line* Q, F   46
honest-true] DYCE; ~, ~ ROWE   48 th'endeavour] Q; the indeauor F   49 Padua]
THEOBALD; Mantua Q, F   50 cousin's hands] Q2; cosin hands Q; cosins hand F   53
traject] ROWE; Tranect Q, F

37 **people** i.e. of her household
46 **honest-true** upright and faithful
(Schmidt)
49 **Padua** See Collation. Not Mantua but
Padua, which Shakespeare later refers
to as Bellario's home (4.1.108, 118;
5.1.268), was the seat of learning for
civil law in Italy (Theobald).
51 **look what** i.e. whatever
52 **imagined speed** Either 'imaginable

swiftness' (Abbott 375) or, better, 'speed
of imagination' (Steevens, Schmidt).
Compare *Hamlet* 1.5.29–30: 'with
wings as swift | As meditation.'
53 **traject** See Collation. A misreading, *an*
for *ai* in the MS spelling 'traiect' is easy
(NS; *TC* 326). The word derives from
Italian *traghetto* ('ferry').
54 **trades to** communicates with (from the
mainland)

BALTHASAR

　Madam, I go with all convenient speed.　　　　　*Exit*

PORTIA

　Come on, Nerissa, I have work in hand
　That you yet know not of. We'll see our husbands
　Before they think of us.

NERISSA　　　　　　　Shall they see us?

PORTIA

　They shall, Nerissa, but in such a habit
　That they shall think we are accomplishèd　　　　60
　With that we lack. I'll hold thee any wager,
　When we are both accoutered like young men
　I'll prove the prettier fellow of the two,
　And wear my dagger with the braver grace,
　And speak between the change of man and boy　　65
　With a reed voice, and turn two mincing steps
　Into a manly stride, and speak of frays
　Like a fine bragging youth, and tell quaint lies
　How honourable ladies sought my love,
　Which I denying, they fell sick and died.　　　　70
　I could not do withal. Then I'll repent,
　And wish for all that that I had not killed them;
　And twenty of these puny lies I'll tell,
　That men shall swear I have discontinued school　　75
　Above a twelvemonth. I have within my mind
　A thousand raw tricks of these bragging Jacks
　Which I will practise.

NERISSA　　　　　　　Why, shall we turn to men?

PORTIA

　Fie, what a question's that,
　If thou wert near a lewd interpreter!　　　　　80

56 *Exit*] Q2; *not in* Q, F　59 us.] Q2; ~? Q, F　80 near] F3; nere Q, F

56 **convenient** appropriate, fit
60 **habit** costume, apparel
61–2 **we are . . . lack** (a bawdy joke)
61 **accomplishèd** equipped, furnished
63 **accoutered** dressed
65 **braver** finer
66–7 **speak . . . voice** Portia means to imitate an adolescent's broken voice.
68 **frays** (1) fights, (2) deflowerings (*OED v.*² 3) (Rubinstein)
69 **quaint** ingeniously elaborated (Onions),

with a possible bawdy allusion to 'female genitals' (Rubinstein); compare *frays* (68).
72 **I . . . withal** I could not help it (Gifford, cited by Furness), but with a bawdy quibble on *do* = 'copulate'.
77 **raw** immature
　Jacks fellows, knaves
78 **turn to** become; but Portia (79–80) quibbles on 'become coitally available to' (Colman). Cf. 1.3.78.

183

But come, I'll tell thee all my whole device
When I am in my coach, which stays for us
At the park gate; and therefore haste away,
For we must measure twenty miles today.   *Exeunt*

3.5   *Enter Lancelot the clown and Jessica*

LANCELOT   Yes, truly; for, look you, the sins of the father
are to be laid upon the children, therefore I promise
you I fear you. I was always plain with you, and so
now I speak my agitation of the matter, therefore be o'
good cheer, for truly I think you are damned. There is   5
but one hope in it that can do you any good, and that
is but a kind of bastard hope, neither.

JESSICA   And what hope is that, I pray thee?

LANCELOT   Marry, you may partly hope that your father
got you not, that you are not the Jew's daughter.   10

JESSICA   That were a kind of bastard hope indeed! So the
sins of my mother should be visited upon me.

LANCELOT   Truly then I fear you are damned both by
father and mother. Thus when I shun Scylla, your
father, I fall into Charybdis, your mother. Well, you   15
are gone both ways.

JESSICA   I shall be saved by my husband; he hath made
me a Christian.

[Handwritten marginal notes: "Jesica was only Jewsh because her father was Jewy it isn't her fault"; "You should crich that you are adopted"; "She is saved because married Lorenzo"]

8 1 my] Q2, F; my my Q
   3.5] *after* Capell; *not in* Q, F   0.1 *Lancelot the*] ROWE (*subs.*); *not in* Q, F   1 LANCELOT] ROWE (*throughout scene*); *Clowne* Q, F   4 o'] CAPELL; a Q; of F   18 Christian.] Q2, F; ~? Q

8 1 **device** plan, scheme
8 2 **coach** Only the rich could afford their own coaches.
3.5.1–2 **sins . . . children** See Exod. 20: 5, 34: 7; Deut. 5: 9. In the Book of Common Prayer and the Catechism, 'sin' is used instead of the biblical 'iniquity' (Noble).
3 **I fear you** i.e. I fear for you (Abbott 200)
4 **agitation** i.e. cogitation (Eccles, cited by Furness), though the malapropism is ironically appropriate
7 **bastard hope** 'i.e. without a true cause to beget it', but leading to the word-play of 9–11 (NCS).
   neither (an intensive)
10 **got** begot

14–15 **Scylla . . . Charybdis** proverbial for being caught in a dilemma (Tilley S169; Dent, 206). In one of his adventures Odysseus had to steer between the monster Scylla and the whirlpool of Charybdis in a narrow sea to avoid catastrophe (*Odyssey* 12. 235–63.). If 'Scylla' was spoken with a short *i* sound, a pun 'Scylla'–'Shylock' is likely (see 'Sources', above).
15 **fall into** (1) drop into, (2) enter sexually (NCS)
16 **gone** i.e. damned
17 **saved . . . husband** Cf. 1 Cor. 7: 14: 'the unbelieving wife is sanctified by the husband' (Noble) and its gloss: 'Meaning, that the faith of the believer

LANCELOT Truly, the more to blame he! We were Christians enough before, e'en as many as could well live,   20
one by another. This making of Christians will raise
the price of hogs. If we grow all to be pork-eaters, we
shall not shortly have a rasher on the coals for money.

    Enter Lorenzo

JESSICA I'll tell my husband, Lancelot, what you say.
Here he comes.   25

LORENZO I shall grow jealous of you shortly, Lancelot, if
you thus get my wife into corners.

JESSICA Nay, you need not fear us, Lorenzo. Lancelot
and I are out. He tells me flatly there's no mercy for me
in heaven because I am a Jew's daughter, and he says   30
you are no good member of the commonwealth, for in
converting Jews to Christians, you raise the price of
pork.

LORENZO (*to Lancelot*) I shall answer that better to the
commonwealth than you can the getting up of the   35
Negro's belly. The Moor is with child by you, Lancelot!

LANCELOT It is much that the Moor should be more than
reason, but if she be less than an honest woman, she is
indeed more than I took her for.

LORENZO How every fool can play upon the word! I   40
think the best grace of wit will shortly turn into

20 e'en] Q2, F; in Q   25 comes.] Q2, F; come? Q   27 corners.] Q2; ~? Q, F   29
there's] Q; there is F   34 *to Lancelot*] OXFORD; *not in* Q, F   36 Lancelot!] ~? Q, F

    hath more power to sanctify marriage
than the wickedness of the other to
pollute it.'

21 **one by another** (1) alongside each
other, (2) off each other

21–3 **This . . . money** Lancelot refers to
the Jewish prohibition against eating
pork (see 1.3.31–5 and 31–3 below).
One function of Lancelot's comedy is
'repeatedly to twist spiritual truths towards the carnal' (Danson, 97).

23 **rasher** i.e. of bacon
for money i.e. for any money

27 **corners** secluded places (*OED* 6)

29 **are out** (1) out in the open, (2) have
quarrelled

35–6 **getting . . . belly** making the Negress
pregnant, with a quibble on 'raise'
(32)

36 **Moor** *Negro* and *Moor* were practically
synonymous through the seventeenth
century: see *OED sb.*[2] *Moor*; compare
Blackamoor.

37–9 **Moor . . . for** Lancelot puns on
'Moor'–'more' (twice) (Kökeritz, 130,
239; compare Cercignani, 191–2),
with quibbles on 'much' (37) and 'less'
(38). Cf. *Titus* 4.2.52–3: 'NURSE . . .
did you see Aaron the Moor? | AARON
Well, more or less, or ne'er a whit at
all' (Staunton, cited by Furness).

37–8 **more than reason** larger than is
reasonable (Onions)

38 **less** Lancelot of course means 'more',
but gets tangled up in his own word-
play.
honest chaste

41 **best grace** most becoming form (Kittredge)

silence, and discourse grow commendable in none
only but parrots. Go in, sirrah, bid them prepare for
dinner.

LANCELOT That is done, sir, they have all stomachs.   45

LORENZO Goodly Lord, what a wit-snapper are you!
Then bid them prepare dinner.

LANCELOT That is done too, sir; only 'cover' is the word.

LORENZO Will you cover then, sir?

LANCELOT Not so, sir, neither. I know my duty.   50

LORENZO Yet more quarrelling with occasion! Wilt thou
show the whole wealth of thy wit in an instant? I pray
thee understand a plain man in his plain meaning: go
to thy fellows, bid them cover the table, serve in the
meat, and we will come in to dinner.   55

LANCELOT For the table, sir, it shall be served in; for the
meat, sir, it shall be covered; for your coming in to
dinner, sir, why, let it be as humours and conceits
shall govern.   *Exit*

LORENZO

O dear discretion, how his words are suited!   60
The fool hath planted in his memory
An army of good words, and I do know
A many fools that stand in better place,
Garnished like him, that for a tricksy word
Defy the matter. How cheer'st thou, Jessica?   65
And now, good sweet, say thy opinion:
How dost thou like the Lord Bassanio's wife?

44 dinner.] ~? Q1–2, F   45 stomachs.] Q2; ~? Q, F   47 dinner.] Q2, F; ~? Q   59
Exit] Q, F (*Exit Clowne.*)   65 cheer'st] Q, F; far'st Q2

45 **stomachs** with a play on 'appetites'

48 **cover** spread the tablecloth, but later (50) Lancelot quibbles on the sense 'cover one's head' (see 2.9.43 n.)

51 **quarrelling with occasion** i.e. Lancelot abuses the opportunity to quibble (Kittredge)

57 **covered** i.e. served in a covered dish

58 **humours and conceits** inclinations and fancies

60 **discretion** discrimination
suited made suitable, adapted

63 **A many** i.e. many; compare *As You Like It* 1.1.110: 'a many merry men with him' (Abbott 87)
stand . . . place i.e. are employed in loftier positions

64 **Garnished** (1) (of clothes) dressed, decked out, cf. 2.6.45; (2) (of language) furnished, cf. *L.L.L.* 2.1.78: 'every one her own hath garnishèd | With such bedecking ornaments of praise.'
tricksy ambiguous (NCS). Lorenzo's comments, like Lancelot's on the 'making of Christians' (21), seem to anticipate events in 4.1; but see Introduction, p. 34.

65 **Defy the matter** i.e. oppose the substance of sense
How cheer'st thou how is it with you

JESSICA

Past all expressing. It is very meet
The Lord Bassanio live an upright life,
For, having such a blessing in his lady,   70
He finds the joys of heaven here on earth,
And if on earth he do not merit it,
In reason he should never come to heaven.
Why, if two gods should play some heavenly match
And on the wager lay two earthly women,   75
And Portia one, there must be something else
Pawned with the other, for the poor rude world
Hath not her fellow.

LORENZO   Even such a husband
Hast thou of me as she is for a wife.

JESSICA

Nay, but ask my opinion too of that.   80

LORENZO

I will anon. First let us go to dinner.

JESSICA

Nay, let me praise you while I have a stomach.

LORENZO

No, pray thee, let it serve for table-talk.
Then, howsome'er thou speak'st, 'mong other things
I shall digest it.

JESSICA   Well, I'll set you forth.   *Exeunt*   85

72–3 merit it, | In] POPE; meane it, it | in Q; meane it, then | In Q2; meane it, it | Is
F   73 heaven.] Q2; ~? Q, F   79 for a] F; for Q   80 that.] Q2; ~? Q, F   81 dinner.]
Q2; ~? Q, F   82 stomach.] Q2; ~? Q, F   84 howsome'er] ROWE; how so mere Q;
howsoere Q2; how som ere F   85 it.] Q2; ~? Q, F   *Exeunt*] F; *Exit.* Q

68 **meet** fitting
69 **live** i.e. should live
72–3 **merit it,** | In See Collation. Q2 and
F emend guessingly; Pope's emendation appears correct and is palaeo-
graphically easy: 'merry it' was
probably misread as 'meane it' by
the Q compositor, who then inserted
the comma between the two pronouns
(see NS; *TC* 326). Jessica says that,
having received in Portia heaven's joys
on earth, Bassanio should live an upright life; otherwise, he will not get
them later.

73 **In reason** i.e. it stands to reason
75 **lay** stake
77 **Pawned** staked
82 **stomach** (1) appetite, (2) inclination
84 **howsome'er** howsoever
85 **digest** (1) put up with, swallow (Onions), (2) assimilate gastronomically,
(3) comprehend, as in *Coriolanus* 1.1.
148, (4) reduce to nothing, as in *All's
Well* 5.3.75
set you forth (1) extol, praise you
greatly, (2) set you up, as for a feast
(Clarendon *subs.*, cited by Brown)

4.1 *Enter the Duke, the magnificoes, Antonio,*
Bassanio, Salerio, and Graziano; ⌈officers and
attendants of the court⌉

DUKE

What, is Antonio here?

ANTONIO   Ready, so please your grace.

DUKE

I am sorry for thee. Thou art come to answer
A stony adversary, an inhuman wretch
Uncapable of pity, void and empty
From any dram of mercy.

ANTONIO   I have heard
Your grace hath ta'en great pains to qualify
His rigorous course, but since he stands obdurate
And that no lawful means can carry me
Out of his envy's reach, I do oppose
My patience to his fury, and am armed   10
To suffer with a quietness of spirit
The very tyranny and rage of his.

DUKE

Go one, and call the Jew into the court.

SALERIO

He is ready at the door. He comes, my lord.

4.1] *after* Rowe; *Actus Quartus* F; *not in* Q   0.1 *Salerio*] *not in* Q, F; ~ *and others*
CAMBRIDGE; Salerio, Solanio, *and others* CAPELL   0.2 *officers . . . court*] NS subs.; *not in*
Q, F   1 What . . . grace] OXFORD; *as two lines* Q, F   grace.] Q2; ~? Q, F

4.1.0 **Enter . . . Graziano** They do not
enter all at once, but the Duke and the
magnificoes probably entered from one
door and then Antonio and the others
from another.
 Duke Whether deliberately or not,
Shakespeare has the Duke, or *doge*,
oversee the court, although historically he had not done so for at least
two centuries. But, as in *Dream* or
Measure, it is dramatically more effective as well as efficient to have a single
ruler preside. Besides, the setting resembles the trial of Roderigo Lopez before Essex and the magnificoes of
London, the city fathers (NS).
 magnificoes the magnates of Venice. In
Venice civil or criminal court proceedings were held before forty judges,
elected from among the nobility (see

T. Elze, *Shakespeare Jahrbuch*, 14 (1879),
178, cited by Furness, Brown). Fewer
magnificoes here are used mainly to
swell a progress or at most suggest a
panel of judges.
 1 **Ready** (1) Here!, (2) Well prepared
(for the inevitable)
 2 **answer** i.e. answer to, satisfy
 3 **stony adversary** 'The scene opens with
blatant partiality' (Merchant).
 5 **From** i.e. of
  dram smallest amount; lit., $\frac{1}{16}$ avoir-
dupois ounce
 6 **qualify** moderate
 9 **envy's** malice's, as in *Richard II*
1.2.21.
 10 **patience** the virtue invariably advocated in the Renaissance during a time
of adversity; cf. *Lear* 1.4.240, 2.2.445
 12 **tyranny** cruelty

Enter Shylock

DUKE

Make room, and let him stand before our face. 15
Shylock, the world thinks—and I think so too—
That thou but leadest this fashion of thy malice
To the last hour of act, and then 'tis thought
Thou'lt show thy mercy and remorse more strange
Than is thy strange apparent cruelty. 20
And where thou now exacts the penalty,
Which is a pound of this poor merchant's flesh,
Thou wilt not only loose the forfeiture,
But, touched with human gentleness and love,
Forgive a moiety of the principal, 25
Glancing an eye of pity on his losses
That have of late so huddled on his back,
Enough to press a royal merchant down
And pluck commiseration of his state
From brassy bosoms and rough hearts of flint, 30
From stubborn Turks and Tartars never trained
To offices of tender courtesy.
We all expect a gentle answer, Jew.

21 exacts] Q; exact'st F   23 loose] Q, F; lose F4, ROWE   29 his state] Q2, F; this states
Q   30 flint] Q2; flints Q, F   33 Jew.] Q2; ~? Q, F

14.1 Unless they are already prepared on a table, Shylock enters carrying his scales and a knife. Neither Tubal nor any other Jew accompanies him. His isolation is dramatically effective (see Granville-Barker, 100 n. 1) and may be further significant—suggesting no support from fellow Jews—although it has not always been staged this way. The delayed entrance allows time for sad expectation to build, and for a hush to fall over the stage audience when Shylock finally stands before the Duke. As the next line indicates, he pushes his way through a crowd of Antonio's supporters. Booth's prompt-book indicates that he entered slowly, bowing to the Duke and showing great deference to him throughout the scene, but to no one else—except Portia, as long as she appeared to favour his suit (Furness).

17 **fashion** mere form, pretence (Onions)
18 **To . . . act** i.e. until the eleventh hour
19 **remorse** pity

19–20 **strange . . . strange** wonderful . . . extraordinary, or 'abnormal', with a hint of 'alien' (NCS)
20 **apparent** i.e. seeming, not real (Johnson)
23 **loose** surrender, but with a possible play on *lose* = 'forget'. The words were not clearly differentiated in Elizabethan English.
25 **moiety** portion. The Duke obviously has no conception of Shylock's state of mind in expecting forgiveness not only of the penalty, but of part of the principal as well.
28 **royal merchant** See 3.2.237 n.
31 **Turks** classed with Jews, Infidels, and Heretics in the Good Friday Collect (Merchant)
Tartars brutal and warlike people from central Asia, led by Genghis Khan in the thirteenth century
32 **offices** duties
33 **gentle** probably punning on *Gentile*, as at 2.4.34.

SHYLOCK

I have possessed your grace of what I purpose,
And by our holy Sabbath have I sworn   35
To have the due and forfeit of my bond.
If you deny it, let the danger light
Upon your charter and your city's freedom.
You'll ask me why I rather choose to have
A weight of carrion flesh than to receive   40
Three thousand ducats. I'll not answer that,
But say it is my humour. Is it answered?
What if my house be troubled with a rat
And I be pleased to give ten thousand ducats
To have it baned? What, are you answered yet?   45
Some men there are love not a gaping pig,
Some that are mad if they behold a cat,
And others when the bagpipe sings i'th'nose
Cannot contain their urine; for affection,
Mistress of passion, sways it to the mood   50

34 SHYLOCK] ROWE; *Iewe* Q, F;   ~? Q   38 freedom.] Q2, F;   ~? Q   41 that,] Q2;   ~? Q;   ~: F
46 pig,] ~? Q;   ~: Q2, F   47 cat,] ~? Q;   ~: Q2, F   49 urine;] CAPELL (*conj.* Thirlby);
~∧ Q, F   affection,] CAPELL (*conj.* Thirlby);   ~. Q, F   50 Mistress] CAPELL (*conj.* Thirlby);
Maisters Q; Masters Q2, F

34 **possessed** informed
35 **have I sworn** '[T]o swear, as all Shakespearean tyrants would testify, means
to impose one's will on the world
rather than lend ear to what the world
imports, as love and gentleness [33]
would do' (Oz, 100).
36 **due and forfeit** i.e. forfeit due (hendiadys)
37-8 **let . . . freedom** 'Venice can no
longer be called a free city if it denies
foreigners the rights that its laws secure to them' (Kittredge). But Venice
was an independent state, not an English town that owed its charter and
freedom to a medieval monarch
(Brown).
37 **danger** harm, damage (Onions)
39-41 **You'll . . . that** In fact, the Duke
has not asked, though the question
must be in his and everyone else's
mind (cf. 3.1.48-9). The answer Shylock gives (after his refusal to state his
real motive) is calculated, as Johnson
observed, to 'aggravate the pain of the
enquirer'.
42 **humour** mental disposition controlled

by the balance, or imbalance, of four
main bodily fluids (blood, phlegm,
choler, and bile). See David H. Bishop,
'Shylock's Humour', *SAB* 23 (1948),
174-80.
45 **baned** poisoned
46 **gaping pig** roasted pig with its mouth
open; perhaps a recollection of Thomas Nashe, *Pierce Pennilesse, his Sup-
plication to the Devil* (1592): 'Some
will take on like a madman, if they see
a pig come to the table' (Malone).
48 **sings i'th'nose** i.e. sounds like a nasal
voice
49 **Cannot . . . urine** an unusual reaction, but testified to by stories of the
period and in Jonson's *Everyman in his
Humour* (1616) 4.2.19-22: 'can he
not hold his water, at reading of a
ballad? |—O, no: a rhyme to him, is
worse than cheese, or a bagpipe.'
affection natural instinct on which the
disposition of the mind depends
(Schmidt); cf. 'humour' 42 n., and
Measure 2.4.168: 'by the affection
that now guides me most.'
50 **Mistress of passion** See Collation. NS

190

Of what it likes or loathes. Now, for your answer:
As there is no firm reason to be rendered
Why he cannot abide a gaping pig,
Why he a harmless necessary cat,
Why he a woollen bagpipe, but of force   55
Must yield to such inevitable shame
As to offend, himself being offended;
So can I give no reason, nor I will not,
More than a lodged hate and a certain loathing
I bear Antonio, that I follow thus   60
A losing suit against him. Are you answered?

BASSANIO

This is no answer, thou unfeeling man,
To excuse the current of thy cruelty.

SHYLOCK

I am not bound to please thee with my answers.

BASSANIO

Do all men kill the things they do not love?   65

SHYLOCK

Hates any man the thing he would not kill?

BASSANIO

Every offence is not a hate at first.

SHYLOCK

What, wouldst thou have a serpent sting thee twice?

ANTONIO

I pray you, think you question with the Jew.
You may as well go stand upon the beach   70

61 him.] ~? Q, F; ~; Q2   63 cruelty.] Q2, F; ~? Q   64, 66, 68 SHYLOCK] Q2; *Iewe*
Q, F   64 answers.] Q, F; ~? Q   67 first.] Q2, F; ~? Q

speculates that copy read 'Mrs.', an
abbreviation for either 'Masters' or
'Mistress'. NCS follows Q/F, emending
'of' to 'oft', resulting in an odd
rhythm untypical even of Shylock; but
see the Supplementary Note, p. 167.

54 **necessary** 'Inevitably determined or
fixed by predestination' (*OED* 5; cited
by Bishop, 42 n., above); compare *2
Henry IV* 3.1.82: 'the necessary form
of this.'

55 **woollen** The 'bags' of the pipes are
commonly wrapped in baize or flannel
(NS).

of force perforce
59 **lodged** deep-seated
   certain fixed, steadfast
61 **losing** unprofitable, because he will
   lose 3,000 ducats if his suit is upheld
63 **current** unimpeded course or progress
   (Onions). The metaphor may derive
   from the conception of a humour
   (black bile) flowing through Shylock's
   body (NCS).
64 **I . . . answers** Bassanio has inter-
   rupted; it was the Duke who put the
   question to Shylock.
69 **question** debate, dispute

And bid the main flood bate his usual height;
You may as well use question with the wolf
Why he hath made the ewe bleat for the lamb;
You may as well forbid the mountain pines
To wag their high tops and to make no noise   75
When they are fretten with the gusts of heaven;
You may as well do anything most hard
As seek to soften that—than which what's harder?—
His Jewish heart. Therefore, I do beseech you,
Make no more offers, use no farther means,   80
But with all brief and plain conveniency
Let me have judgement and the Jew his will.

BASSANIO (*to Shylock*)
For thy three thousand ducats here is six.

SHYLOCK
If every ducat in six thousand ducats
Were in six parts, and every part a ducat,   85
I would not draw them. I would have my bond.

DUKE
How shalt thou hope for mercy, rend'ring none?

SHYLOCK
What judgement shall I dread, doing no wrong?
You have among you many a purchased slave
Which, like your asses and your dogs and mules,   90
You use in abject and in slavish parts,

---

72 You may as] Q *corr.*, Q2; Or euen as F; *not in* Q *uncorr.* wolf] Woolfe, Q *corr.*; ~,
Q *uncorr.*, Q2, F   73 Why . . . made] Q *corr.*, Q2; *not in* Q *uncorr.*, F   bleat] F; bleake
Q   74 mountain] F; ~ of Q   76 fretten] Q; fretted F   78 what's] Q; what F   79
heart.] F; ~? Q, Q2   80 more] F; moe Q   82 will.] Q2, F; ~? Q   83 six.] Q2, F; ~?
Q   84, 88 SHYLOCK] ROWE; *Iewe* Q, F   86 bond.] Q2; ~? Q, F

71 **main flood** high tide
   bate abate, decrease
72 **use question with** enquire of
73 **bleat** See Collation. Although Q's
   bleake (pronounced like *blake*) may be
   dialectal for *bleat*, it is more likely an
   error, corrected by F but not Q2 (*TC*
   326).
76 **fretten** archaic form of *fretted*
81 **conveniency** fitness, propriety
86 **draw** take, receive.
87 **How . . . none** Cf. James 2: 13: 'For
   there shall be judgement merciless to
   him that showeth no mercy', and
   Matt. 5: 7: 'Blessed are the merciful:

for they shall obtain mercy' (Shaheen).
88 **What . . . wrong** Despite his moral
   obtuseness, Shylock is sincere in so far
   as he believes the law is on his side.
   Nevertheless, the arguments that follow hit home against Christian hypocrisy, as critics (e.g. Johnson) see it.
89 **purchased slave** Not only in Venice,
   but in all of Europe, including England, slavery was practised. In the
   1987 RSC production Shylock seized a
   black attendant at this point and held
   him forward during this speech.
91 **parts** duties, tasks

Because you bought them. Shall I say to you,
'Let them be free, marry them to your heirs.
Why sweat they under burdens? Let their beds
Be made as soft as yours, and let their palates   95
Be seasoned with such viands'? You will answer
'The slaves are ours'. So do I answer you.
The pound of flesh which I demand of him
Is dearly bought: 'tis mine, and I will have it.
If you deny me, fie upon your law!   100
There is no force in the decrees of Venice.
I stand for judgement. Answer: shall I have it?

DUKE

Upon my power I may dismiss this court,
Unless Bellario, a learned doctor
Whom I have sent for to determine this,   105
Come here today.

SALERIO   My lord, here stays without
A messenger with letters from the doctor,
New come from Padua.

DUKE

Bring us the letters! Call the messenger!

                                            Exit Salerio

BASSANIO

Good cheer, Antonio! What, man, courage yet!   110
The Jew shall have my flesh, blood, bones, and all
Ere thou shalt lose for me one drop of blood.

ANTONIO

I am a tainted wether of the flock,
Meetest for death. The weakest kind of fruit
Drops earliest to the ground; and so let me.   115
You cannot better be employed, Bassanio,

99 'tis] Q2, F; as Q   106 today.] Q2, F; ~? Q   108 Padua.] Q2, F; ~? Q   109 letters!]
~? Q; Letters, Q2, F   messenger!] ~? Q; ~. Q2; Messengers. F   *Exit Salerio*] OXFORD;
not in Q, F   110 Antonio! . . . yet!] ~? . . . : Q; ~, . . . ~: Q2; ~. . . . ~: F   112
blood.] Q2, F; ~? Q

102 **stand for** (1) make a stand for
  (Onions), (2) represent, as at 141 and
  3.2.57 (Schmidt; cf. Lewalski, 337–
  8), (3) stand in hope of
103 **Upon** i.e. in accordance with
106 **stays without** waits outside
112 **one drop of blood** See Introduction,
  p. 49.

113 **tainted** diseased, with a suggestion of
  moral degeneration or corruption;
  perhaps a sacrificial scapegoat
  wether castrated ram. Abraham substituted a ram for the offering of his
  son Isaac (Gen. 22: 13). See Introduction, p. 49 n. 1.

Than to live still and write mine epitaph.

    Enter ⌈Salerio, with⌉ Nerissa dressed as a lawyer's
    clerk

DUKE

Came you from Padua, from Bellario?

NERISSA

From both, my lord. (*Presenting a letter*) Bellario greets
    your grace.

    Shylock whets his knife on his shoe

BASSANIO

Why dost thou whet thy knife so earnestly?   120

SHYLOCK

To cut the forfeiture from that bankrupt there.

GRAZIANO

Not on thy sole, but on thy soul, harsh Jew,

Thou mak'st thy knife keen. But no metal can—

No, not the hangman's axe—bear half the keenness

Of thy sharp envy. Can no prayers pierce thee?   125

SHYLOCK

No, none that thou hast wit enough to make.

GRAZIANO

O, be thou damned, inexecrable dog,

And for thy life let justice be accused!

Thou almost mak'st me waver in my faith

To hold opinion with Pythagoras,   130

---

117 epitaph.] Q2, F; ~? Q   117.1 Salerio, with] OXFORD; *not in* Q, F   *dressed . . . clerk*] ROWE (*subs.*); *not in* Q, F   119 From . . . grace] Q; *two lines breaking after* 'both' F   both, my lord.] Q2 (~, ~L.); both? my L. Q; ~. | My Lord‸ F   *Presenting a letter*] CAPELL (*subs., following* Rowe); *not in* Q, F   grace.] Q2, F; ~? Q   119.1] OXFORD; *not in* Q, F   121, 126, 138 SHYLOCK] ROWE; Iewe Q, F   121 forfeiture] Q, F; forfeit ROWE   there.] F; ~? Q; ~, Q2   128 accused!] NCS; ~; Q, Q2; ~: F

117 **to live . . . epitaph** Compare Hamlet's dying wish to Horatio, 5.2.300–1.

119.2 The stage direction is indicated in the text, 120, 122–3.

122 **sole . . . soul** a common pun, hardly calculated to raise a laugh here (Kittredge); also in *Romeo* 1.4.14–16

124 **hangman's** executioner's
keenness (1) sharpness, (2) savagery (Bevington; *OED keen adj. 2c*)

126 **wit** intelligence, cleverness

127 **inexecrable** that cannot be cursed enough (Onions). Emended to 'inexorable' (F3), 'inexecrable' though

unique in Shakespeare (and perhaps in the language; *TC* 326) appears as an intensive of *execrable* (Malone). Cf. Marlowe's *Dr Faustus*: 'thou damned witch and execrable dog' (Merchant).

128 **for . . . accused** i.e. what an injustice to humanity it was that you were ever born

129–32 **Thou . . . men** Graziano refers to the Pythagorean doctrine of the transmigration of souls, a heresy to Christians. As the false curate Sir Topas, Feste comically examines Malvolio on the doctrine, *Twelfth Night* 4.2.50–60.

That souls of animals infuse themselves
Into the trunks of men. Thy currish spirit
Governed a wolf who, hanged for human slaughter,
Even from the gallows did his fell soul fleet,
And whilst thou layest in thy unhallowed dam   135
Infused itself in thee; for thy desires
Are wolvish, bloody, starved, and ravenous.

SHYLOCK

Till thou canst rail the seal from off my bond
Thou but offend'st thy lungs to speak so loud.
Repair thy wit, good youth, or it will fall   140
To cureless ruin. I stand here for law.

DUKE

This letter from Bellario doth commend
A young and learnèd doctor to our court.
Where is he?

NERISSA   He attendeth here hard by,
To know your answer, whether you'll admit him.   145

DUKE

With all my heart. Some three or four of you
Go give him courteous conduct to this place.

Exeunt three or four

Meantime the court shall hear Bellario's letter.

(*Reads*) Your grace shall understand that at the receipt of
your letter I am very sick, but in the instant that your   150
messenger came, in loving visitation was with me a

---

141 cureless] Q; endlesse F   143 to] Q; in F   147.1 *Exeunt . . . four*] OXFORD; *Exeunt
officials* NCS; *not in* Q, F   149 *Reads*] CAPELL, *who however adds speech prefix 'Cle[rk]': not
in* Q, F; SISSON *assigns speech to Nerissa*

---

133 **wolf** 'Usurers were often called
wolves; the wolf is an emblem of Envy
in sixteenth-century literature; and
predatory animals were occasionally
tried and executed in Europe till the
late seventeenth century' (NCS).
who . . . slaughter a nominative absolute construction, used by Elizabethan
authors to emphasize the object which
is superfluously repeated (Abbott 376)
134 **fell** cruel
fleet flit, pass from the body (*OED* 10*b*)
135 **dam** Graziano extends the animal
image to Shylock's mother.
137 **starved** hungry, famished; contextu-

ally and etymologically, perhaps,
'deadly' (from OE *steorfan* 'to die'; Q/F
steru'd)
138 **rail** declaim abusively
139 **but offend'st** only harm
140 **fall** i.e. as a house falls
151 **in loving visitation** i.e. on a friendly
visit. That both the Duke and Portia
should seek assistance from the
learned Bellario, presumably the fore-
most legal authority in the region, is
hardly coincidental. The letter's state-
ments to the contrary, however, no
visit between Portia and Bellario could
have occurred, since Portia instructs

Portia's alter ego

young doctor of Rome; his name is Balthasar. I acquainted him with the cause in controversy between
the Jew and Antonio, the merchant. We turned o'er
many books together. He is furnished with my opinion   155
which, bettered with his own learning—the greatness
whereof I cannot enough commend—comes with him
at my importunity to fill up your grace's request in my
stead. I beseech you, let his lack of years be no impedi-
ment to let him lack a reverend estimation, for I never   160
knew so young a body with so old a head. I leave him
to your gracious acceptance, whose trial shall better
publish his commendation.

> *Enter ⌈three or four with⌉ Portia as Balthasar*
> *dressed as a Doctor of Laws*

You hear the learned Bellario, what he writes,
And here, I take it, is the doctor come.   165
Give me your hand. Come you from old Bellario?

PORTIA
I did, my lord.

DUKE   You are welcome. Take your place.
Are you acquainted with the difference
That holds this present question in the court?

PORTIA
I am informèd throughly of the cause.   170
Which is the merchant here, and which the Jew?

DUKE
Antonio and old Shylock, both stand forth.

PORTIA
Is your name Shylock? *Would be obvious but Portia*
chose to ask anyway.

163.1 *three . . . with*] OXFORD; *not in* Q, F   *as*] OXFORD; *for* Q, F   *dressed . . . Laws*]
ROWE (*subs.*); *not in* Q, F   164 You] NS; *Duke.* You Q, F   166 Come] Q; Came F

her servant to bring notes and garments from Bellario in Padua to her in
Venice (NS). It is this Balthasar who
visits Dr Bellario (3.4.45–55).

160 **let him lack** i.e. deprive him of
162–3 **whose . . . commendation** i.e. the
test you make of him will better proclaim his ability (than these words)
167 **your place** i.e. beside me? among
the *magnificoes*? However Shakespeare

staged the scene, most actresses today
freely move about the stage among
Shylock and the others.
168 **difference** dispute
170 **throughly** thoroughly
cause i.e. of the proceedings
171 **Which . . . Jew** an ingenuous question, if their costumes easily distinguish the two (see 1.3.109 n.);
however, ingenuousness may be part
of Portia's initial gambit.

SHYLOCK   Shylock is my name.

PORTIA

Of a strange nature is the suit you follow,

Yet in such rule that the Venetian law   175

Cannot impugn you as you do proceed.

(*To Antonio*) You stand within his danger, do you not?

ANTONIO

Ay, so he says.

PORTIA   Do you confess the bond?

ANTONIO

I do.

PORTIA Then must the Jew be merciful.

SHYLOCK

On what compulsion must I? Tell me that.   180

PORTIA

The quality of mercy is not strained.

It droppeth as the gentle rain from heaven

Upon the place beneath. It is twice blest:

It blesseth him that gives and him that takes.

'Tis mightiest in the mightiest. It becomes   185

The thronèd monarch better than his crown.

His sceptre shows the force of temporal power,

The attribute to awe and majesty,

173 SHYLOCK] ROWE; *Iew* Q, F   180 SHYLOCK] Q; *Iew* F

175 **rule** proper order
176 **impugn** oppose, controvert
177 **danger** (1) debt, (2) power to harm (Onions)
179–80 **must . . . must** Portia's 'must' implies a moral or humane imperative (i.e. since the bond is confirmed by the law and conceded by the defendant, only the Jew's mercy can save Antonio); Shylock's 'must' implies coercion and leads directly into Portia's definition of unconstrained mercy.
181–99 **The quality . . . mercy** A set speech, among the most memorable in all of Shakespeare, Portia's disquisition on mercy recalls debates on the conflicting claims of mercy and justice in the Renaissance and earlier, e.g. Seneca's *De clementia* 1.19 (cited by NS) and Alexandre Silvayn's *The Orator* (cited by Merchant; see 'Sources',

above). For Shakespeare's continuing interest in the debate see e.g. *As You Like It* 4.3.120–38, Isabella's pleas in *Measure* 2.2, and *Tempest* 5.1.11–32. Speaking in 'the accents of Belmont', Portia addresses an 'uncomprehending listener' who is 'resolutely a creature of the earth' (Leggatt, 137).
182 **gentle rain** Cf. Ecclus. 35: 19: 'Oh, how fair a thing is mercy in time of anguish and trouble! It is like a cloud of rain, that cometh in the time of a drought'; and Deut. 32: 2: 'My doctrine shall drop as the rain, and my speech shall distil as doth the dew, as the shower upon the herbs, and as the great rain upon the grass' (Noble).
188 **awe and majesty** possibly hendiadys, i.e. awful majesty (Allen, cited by Furness)

Wherein doth sit the dread and fear of kings;
But mercy is above this sceptred sway.   190
It is enthronèd in the hearts of kings;
It is an attribute to God himself,
And earthly power doth then show likest God's
When mercy seasons justice. Therefore, Jew,
Though justice be thy plea, consider this: 195
That in the course of justice none of us
Should see salvation. We do pray for mercy,
And that same prayer doth teach us all to render
The deeds of mercy. I have spoke thus much
To mitigate the justice of thy plea,   200
Which if thou follow, this strict court of Venice
Must needs give sentence 'gainst the merchant there.

SHYLOCK

My deeds upon my head! I crave the law,
The penalty and forfeit of my bond.

PORTIA

Is he not able to discharge the money?   205

---

192 **It is . . . himself** God's mercy is a dominant theme in the communion service, litany, morning service, etc. (Shaheen) and throughout the Hebrew prayer book, which incorporates e.g. Ps. 136 and its refrain: 'for his mercy endureth for ever'.

193–4 **earthly . . . justice** a commonplace (Tilley M898), as in *Titus* 1.1.117–18 and *Edward III* (1596): 'And kings approach the nearest unto God, | By giving life and safety unto men' (Malone). Compare Ps. 102: 10–11: 'He hath not dealt with us after our sins, nor rewarded us according to our iniquities. For as high as the heaven is above the earth, so great is his mercy toward them that fear him' (Noble).

194 **seasons** tempers

195–7 **Though . . . salvation** Cf. Ps. 143: 2, quoted at morning prayer (Noble), and the Geneva Bible gloss: 'in God's sight all men are sinners.'

197–9 **We . . . mercy** Portia alludes to the Lord's Prayer, perhaps inappropriately addressing a Jew, but not Shakespeare's Christian audience. On the other hand, both Jew and Gentile might recall Ecclus 28: 2: 'Forgive thy neighbour the hurt that he hath done to thee, so shall thy sins be forgiven thee also, when thou prayest' (Noble). Portia, of course, has not witnessed Shylock's speech, 3.3.13–16.

200 **mitigate** moderate, lessen

203 **My . . . head** An echo, perhaps, of Matt. 27: 25 (Henley, cited by Malone). The Geneva Bible glosses: 'If [Jesus'] death be not lawful, let the punishment fall on our [i.e. the Jews'] heads and our children's.' More immediately, Shylock echoes 'deeds of mercy' (199) (NS).

205–7 **Is he . . . sum** Shylock is standing firm on a technicality, i.e. that Antonio has missed his day. Actually, by Shakespeare's time an appeal in equity against an 'intolerable forfeiture' was usually set aside and an equitable 'penalty' (here the return of principal with interest) would be substituted for the forfeit (Merchant). However cognizant Shakespeare was of this practice, he was intent here upon a certain moral issue and dramatic effect.

BASSANIO

Yes, here I tender it for him in the court,
Yea, twice the sum. If that will not suffice,
I will be bound to pay it ten times o'er
On forfeit of my hands, my head, my heart.
If this will not suffice, it must appear   210
That malice bears down truth. And, I beseech you,
Wrest once the law to your authority.
To do a great right, do a little wrong,
And curb this cruel devil of his will.

PORTIA

It must not be. There is no power in Venice   215
Can alter a decree establishèd.
'Twill be recorded for a precedent,
And many an error by the same example
Will rush into the state. It cannot be.

SHYLOCK

A Daniel come to judgement, yea, a Daniel!   220
O wise young judge, how I do honour thee!

PORTIA

I pray you, let me look upon the bond.

SHYLOCK

Here 'tis, most reverend doctor, here it is.

220, 223, 232, 243 SHYLOCK] Q; *Iew* F   221 I do] Q; do I F

211 **bears down** overwhelms, overthrows
(Onions)
  truth variously glossed as 'reason; the
  reasonable offers of accommodation'
  (Theobald), 'rule of equity' (Heath, *Revisal of Shakespeare's Text* (1765); cited
  by Furness), 'honesty' (Johnson), 'righteousness' (Riverside)
212 **once** i.e. for once
213 **To do ... wrong** Cf. 'The end justifies
  the means' (Tilley E112; Dent, 100).
214 **will** (1) desire, (2) lustfulness
215-16 **no power . . . establishèd** 'Laws
  in Venice had the immutability attributed to the Medes and Persians'
  (Merchant).
217 **precedent** English law is essentially
  case-law, established by precedent or
  precedents, not a code (see previous
  note). Shakespeare here apparently
  mingles the two kinds.
218 **error** i.e. political, not moral error;
  a departure from constitutional practice (NS, paraphrasing Schmidt)
220 **A Daniel** Shylock alludes to the story
  of Susanna and the Elders in the Apocrypha. Daniel, like Portia, was a youth
  but with the wisdom of an elder (see
  248 below). Unlike the Elders who
  bore false witness, he judged righteously; hence, Graziano also gleefully
  uses the name (329), not only because
  events turn about (cf. 'The innocent
  and righteous thou shalt not slay', Susanna 53), but also because Daniel
  convicted the Elders 'by their own
  mouth' (61), as Brown notes. The
  name *Daniel* in Hebrew means 'The
  Judge of the Lord' and was glossed in
  the Geneva Bible as 'The Judgement of
  God'; moreover, in the Book of Daniel
  the prophet Daniel is named 'Baltassar' (Lewalski, 340).
223 **Here 'tis** Shylock eagerly hands the
  bond to Portia.

PORTIA
   Shylock, there's thrice thy money offered thee.

SHYLOCK
   An oath, an oath! I have an oath in heaven.   225
   Shall I lay perjury upon my soul?
   No, not for Venice.

PORTIA   Why, this bond is forfeit.
   And lawfully by this the Jew may claim
   A pound of flesh, to be by him cut off
   Nearest the merchant's heart. (*To Shylock*) Be merciful.   230
   Take thrice thy money. Bid me tear the bond.

SHYLOCK
   When it is paid according to the tenor.
   It doth appear you are a worthy judge;
   You know the law, your exposition
   Hath been most sound. I charge you by the law   235
   Whereof you are a well-deserving pillar,
   Proceed to judgement. By my soul I swear
   There is no power in the tongue of man
   To alter me. I stay here on my bond.

ANTONIO
   Most heartily I do beseech the court
   To give the judgement.   240

PORTIA   Why then, thus it is:
   You must prepare your bosom for his knife.

SHYLOCK
   O noble judge! O excellent young man!

PORTIA
   For the intent and purpose of the law
   Hath full relation to the penalty   245
   Which here appeareth due upon the bond.

SHYLOCK
   'Tis very true. O wise and upright judge!
   How much more elder art thou than thy looks!

227 No] Q2, F; Not Q   232 tenor] Q2; tenure Q, F   247, 249, 253, 256, 259 SHYLOCK]
Q2; *Iew* Q, F

224 **thrice** Bassanio had offered 'twice the
   sum' (207). Either Shakespeare slip-
   ped, the compositor misread copy, or
   Portia (whose money it is) raises the
   offer here and at 231 (NCS).

225 **an oath** See 3.1.122 n.
232 **the tenor** the bond's terms
239 **stay** stand firm
245 **Hath . . . to** i.e. fully authorizes
   (Riverside)

PORTIA *(to Antonio)*
Therefore lay bare your bosom.
SHYLOCK   Ay, his breast.
So says the bond; doth it not, noble judge?   250
'Nearest his heart'—those are the very words.
PORTIA
It is so. Are there balance here to weigh the flesh?
SHYLOCK
I have them ready.
PORTIA
Have by some surgeon, Shylock, on your charge
To stop his wounds, lest he do bleed to death.   255
SHYLOCK
Is it so nominated in the bond?
PORTIA
It is not so expressed, but what of that?
'Twere good you do so much for charity.
SHYLOCK
I cannot find it; 'tis not in the bond.
PORTIA
You, merchant, have you anything to say?   260
ANTONIO
But little. I am armed and well prepared.
Give me your hand, Bassanio; fare you well.
Grieve not that I am fall'n to this for you,
For herein Fortune shows herself more kind
Than is her custom; it is still her use   265
To let the wretched man outlive his wealth,
To view with hollow eye and wrinkled brow
An age of poverty—from which ling'ring penance
Of such misery doth she cut me off.

249 *to Antonio*] OXFORD; *not in* Q, F   255 do] Q; should F   256 Is it so] Q; It is not
F   260 You] Q; Come F

252 **It . . . flesh** an alexandrine, or
hexameter line, as in Q/F. Although
dividing them ('so. | It') encour-
ages a dramatically effective pause
after 'so', alexandrines appear elsewhere in the text (see 1.3.122 n.
above and 317 n.).
balance variant plural form (*OED* 2*b*).
In Shylock's hands the scales cynic-
ally parody the traditional emblem of
Justice.
254 **on your charge** at your expense
256 **nominated** named, stipulated. Here,
Shylock reaches to receive the bond
back from Portia (223 n.).
261 **armed** mentally fortified, steeled
265 **still** always
268 **age** old age

Commend me to your honourable wife.   270
Tell her the process of Antonio's end.
Say how I loved you, speak me fair in death.
And when the tale is told, bid her be judge
Whether Bassanio had not once a love.
Repent but you that you shall lose your friend,   275
And he repents not that he pays your debt;
For if the Jew do cut but deep enough,
I'll pay it instantly with all my heart.

BASSANIO
Antonio, I am married to a wife
Which is as dear to me as life itself;   280
But life itself, my wife, and all the world
Are not with me esteemed above thy life.
I would lose all, ay, sacrifice them all
Here to this devil, to deliver you.

PORTIA (*aside*)
Your wife would give you little thanks for that,   285
If she were by to hear you make the offer.

GRAZIANO
I have a wife who, I protest, I love.
I would she were in heaven, so she could
Entreat some power to change this currish Jew.

NERISSA (*aside*)
'Tis well you offer it behind her back;   290
The wish would make else an unquiet house.

SHYLOCK (*aside*)
These be the Christian husbands. I have a daughter:

275 but] Q; not F   278 instantly] Q, F; presently Q2   292, 298, 301, 315 SHYLOCK]
ROWE; *Iew* Q, F

271 **process** (1) manner, (2) legal proced-
    ure (Merchant)
272 **me . . . death** well of me after I am
    dead
274 **love** i.e. friend, as in Sonnet 13.1
    (Pooler, cited by Brown)
275 **Repent but you** regret only (*OED re-
    pent v.* 2)
278 **with . . . heart** (1) most willingly, (2)
    lit., with my heart (as Shylock wants).
    Compare Mercutio's dying speech,
    Romeo 3.2.96–8, also typical of a
    Renaissance gentleman's witty non-
    chalance, or *sprezzatura*, upon dying

(Merchant, NCS).
280 **Which** 'less definite than *who*. *Who*
    indicates an individual, *which* a "kind
    of person" ' (Abbott 266).
285.1 Portia's speech, like Nerissa's
    (290–1), is properly an aside, since the
    humour is inappropriate to the stage
    audience, however it may be to the
    theatre audience in preparing us for
    the clever trick and its outcome.
292.1 By contrast, the humour in Shy-
    lock's aside is grim, uttered perhaps
    more to himself than to an audience.

Would any of the stock of Barabas
Had been her husband rather than a Christian! —
We trifle time. I pray thee, pursue sentence.   295

PORTIA

A pound of that same merchant's flesh is thine.
The court awards it, and the law doth give it.

SHYLOCK

Most rightful judge!

PORTIA

And you must cut this flesh from off his breast.
The law allows it, and the court awards it.   300

SHYLOCK

Most learnèd judge! A sentence! (*To Antonio*) Come,
    prepare!

PORTIA

Tarry a little; there is something else.
This bond doth give thee here no jot of blood;
The words expressly are 'a pound of flesh'.
Take then thy bond. Take thou thy pound of flesh;   305
But in the cutting it, if thou dost shed
One drop of Christian blood, thy lands and goods
Are by the laws of Venice confiscate
Unto the state of Venice.

GRAZIANO

O upright judge! Mark, Jew. O learnèd judge!   310

SHYLOCK

Is that the law?

305 Take then] Q; Then take F   310 O upright . . . judge] POPE; *lines break* 'Judge, |
Marke' Q, F

293 **Barabas** The murderer chosen for re-
    prieve over Jesus (Luke 23: 18–19),
    hence a figure of revulsion; also the
    name of Marlowe's Jew, accented on
    the first syllable.
295 **trifle** waste
302 **Tarry a little** For the relation of this
    episode to the sacrifice of Isaac, see
    Introduction, p. 49 n. 1.
303 **no . . . blood** Shakespeare again ig-
    nores generally accepted legal princi-
    ples (see 205–7 n.), which held that
    'any right assumes the conditions
    which make the exercise of the right
    possible' (Merchant). The right to take

the pound of flesh would assume the
right to take the blood necessarily
spilled in the process, unless the bond
specifically stipulated that blood
should not be spilled (Haynes, *Outline
of Equity* (1858), 19–20, cited by Furness). But more is involved here than
mere verbal quibbling: Shylock has
repeatedly stood upon the letter of the
law, and Portia here uses the letter to
defeat the letter (see Introduction, pp.
51–2, and Leggatt, 138).
308 **confiscate** confiscated, adjudged for-
    feited (Onions)

PORTIA   Thyself shalt see the act;
  For as thou urgest justice, be assured
  Thou shalt have justice, more than thou desir'st.
GRAZIANO
  O learnèd judge! Mark, Jew—a learnèd judge!
SHYLOCK
  I take this offer, then. Pay the bond thrice   315
  And let the Christian go.
BASSANIO   Here is the money.
PORTIA
  Soft! The Jew shall have all justice. Soft! No haste!
  He shall have nothing but the penalty.
GRAZIANO
  O Jew! An upright judge, a learnèd judge!
PORTIA
  Therefore prepare thee to cut off the flesh.   320
  Shed thou no blood, nor cut thou less nor more
  But just a pound of flesh. If thou tak'st more
  Or less than a just pound, be it but so much
  As makes it light or heavy in the substance
  Or the division of the twentieth part   325
  Of one poor scruple—nay, if the scale do turn
  But in the estimation of a hair,
  Thou diest, and all thy goods are confiscate.
GRAZIANO
  A second Daniel, a Daniel, Jew!
  Now, infidel, I have you on the hip.   330
PORTIA
  Why doth the Jew pause? Take thy forfeiture.

313 desir'st] Q; desirest Q2, F   317 Soft . . . haste] Q, F; CAPELL *divides* 'Soft! | The'
323 but] Q; *not in* F   330 you] Q; thee F

312–13 **For . . . desir'st** Cf. James 2: 13:
    'For there shall be judgement merciless
    to him that showeth no mercy' (Sha-
    heen).
317 **Soft** i.e. wait a moment. Capell's
    lineation (see Collation) emphasizes
    the need for a pause. Cf. Wright, 126,
    who accepts the 'portentous mono-
    syllabic line'.
    all nothing but
323 **just** exact; Cf. 303 n.

324 **substance** gross weight
325–6 **Or . . . scruple** an appositional
    phrase. One twentieth of a scruple
    (about a gram in apothecary's meas-
    ure) is one grain.
326–7 **scale . . . hair** i.e. either the scale's
    indicator is moved by a hair's breadth,
    or as little as the weight of a hair tips
    the scale (Clarendon, cited by Furness)
330 **on the hip** See 1.3.43 n.

[handwritten: Shylock is defeated and now only wants his money back]

SHYLOCK

Give me my principal, and let me go.

BASSANIO

I have it ready for thee. Here it is.

PORTIA

He hath refused it in the open court.

He shall have merely justice and his bond.   335

[handwritten: And here Portia talks away any chance for Shylock to break even out of the deal]

GRAZIANO

A Daniel, still say I, a second Daniel!

I thank thee, Jew, for teaching me that word.

SHYLOCK

Shall I not have barely my principal?

PORTIA

Thou shalt have nothing but the forfeiture

To be so taken at thy peril, Jew.   340

SHYLOCK

Why then, the devil give him good of it!

I'll stay no longer question.

PORTIA   Tarry, Jew,

The law hath yet another hold on you.

It is enacted in the laws of Venice,

If it be proved against an alien   345

That by direct or indirect attempts

He seek the life of any citizen,

The party 'gainst the which he doth contrive

Shall seize one half his goods; the other half

Comes to the privy coffer of the state,   350

And the offender's life lies in the mercy

Of the Duke only, 'gainst all other voice—

In which predicament I say thou stand'st;

For it appears by manifest proceeding

That indirectly, and directly too,   355

[handwritten: Here Portia ruins Shylock's honesty by taking all his money and livelihood.]

340 so taken] Q; taken so F   342 longer] Q, F; ~ heere in Q2   350 coffer] Q, F; coster Q2

335 **merely** (1) only, (2) absolutely (NCS)
342 **I'll . . . question** I'll remain no longer to argue the case. Shylock tries to leave, in disgust at being outwitted and perhaps in fear of what may come next (cf. 313).
345 **alien** See 'Shakespeare and Semitism', pp. 2–3, above.

348 **contrive** plot, scheme
349 **seize** take possession of
350 **privy coffer** personal treasury. In English law, fines were paid to the sovereign.
352 **'gainst . . . voice** i.e. the Duke had absolute power here

Thou hast contrived against the very life
Of the defendant, and thou hast incurred
The danger formerly by me rehearsed.
Down, therefore, and beg mercy of the Duke.

GRAZIANO

Beg that thou mayst have leave to hang thyself!   360
And yet, thy wealth being forfeit to the state,
Thou hast not left the value of a cord;
Therefore thou must be hanged at the state's charge.

DUKE (*to Shylock*)

That thou shalt see the difference of our spirit,
I pardon thee thy life before thou ask it.   365
For half thy wealth, it is Antonio's.
The other half comes to the general state,
Which humbleness may drive unto a fine.

PORTIA

Ay, for the state, not for Antonio.

SHYLOCK

Nay, take my life and all! Pardon not that!   370
You take my house when you do take the prop
That doth sustain my house; you take my life
When you do take the means whereby I live.

PORTIA

What mercy can you render him, Antonio?

GRAZIANO

A halter, gratis. Nothing else, for God's sake!   375

ANTONIO

So please my lord the Duke and all the court
To quit the fine for one half of his goods,

358 formerly] Q2, F; formorly Q; formally HANMER   364 spirit] Q, F; spirits Q2

358 **The danger . . . rehearsed** the penal-
    ties described by me earlier
364 **our** (1) royal plural, (2) Christian
    (NCS)
368 **Which . . . fine** 'if you bear yourself
    humbly, I may be induced to commute
    the forfeit of half your goods to a mere
    fine' (Kittredge).
369 **Ay . . . Antonio** Portia clarifies which
    half the Duke means.
370-3 **Nay . . . live** Shylock's rejoinder
    shows not only how much spirit he
    has left, but also how utterly depend-
    ent he was upon money to exist (see

390 n.). Brown compares *Jew of Malta*
1.2.147–53; Shaheen, Ecclus. 34:
23: 'He that taketh away his neighbour's living, slayeth him', glossed
with a cross-reference to Deut. 24: 14:
'Thou shalt not oppress . . . the stranger
that is in thy land within thy gates.'
374 **What . . . Antonio** The Duke has
already rendered his mercy, unsought
(365). Portia now appeals to Antonio
to show his Christian spirit.
375 **halter** hangman's noose (cf. 360-2)
377 **quit** remit. Antonio refers to the
state's entire half of Shylock's wealth,

I am content, so he will let me have
The other half in use, to render it
Upon his death unto the gentleman   380
That lately stole his daughter.
Two things provided more: that for this favour
He presently become a Christian;
The other, that he do record a gift
Here in the court, of all he dies possessed   385
Unto his son Lorenzo and his daughter.

DUKE ·

He shall do this, or else I do recant
The pardon that I late pronouncèd here.

PORTIA

Art thou contented, Jew? What dost thou say?

SHYLOCK

I am content.

PORTIA (*to Nerissa*) Clerk, draw a deed of gift.   390

SHYLOCK

I pray you, give me leave to go from hence.
I am not well. Send the deed after me,
And I will sign it.

DUKE   Get thee gone, but do it.

GRAZIANO (*to Shylock*)

In christ'ning shalt thou have two godfathers.
Had I been judge, thou shouldst have had ten more—   395

---

394 GRAZIANO] Q2, F; *Shy.* Q   shalt thou] Q; thou shalt F

not just the fine to which it might be
reduced.

379 **in use** i.e in trust, though not for
purposes of usury. Has Antonio heard
how Lorenzo and Jessica have squandered what they stole (3.1.79–105)?

383 **presently** at once
become a Christian For this proviso, so
problematic to modern audiences, see
Introduction, pp. 52–4.

384–6 **The other . . . daughter** This condition refers back to 377. Antonio in-
tends that all of Shylock's fortune,
including the part he holds in trust
(378–81), will eventually revert to
Lorenzo and Jessica.

386 **son** i.e. son-in-law

390 **I am content** Agreement to Antonio's
terms is sometimes taken as further

evidence that Shylock was interested
in money above everything, as in
Patrick Stewart's representation (see
stage history, above).

391–2 **I pray . . . well** Shylock's exit here
is well motivated. He has suffered a
terrible shock, but for him to remain
on stage any longer might swing audience sympathy rapidly towards him
(NCS). Olivier ended his part with a
piercing offstage cry of both physical
and mental anguish.

395 **ten more** i.e. to make up a twelve-
person jury at his trial for conspiring
against Antonio's life. An old joke,
cited by Steevens and Malone, refers
to a jury as 'godfathers'. But the entourage Graziano means may be the
halberdiers or other officers conducting
a criminal to the scaffold (Kittredge).

To bring thee to the gallows, not the font.

Exit Shylock

DUKE

Sir, I entreat you home with me to dinner.

PORTIA

I humbly do desire your grace of pardon.

I must away this night toward Padua,

And it is meet I presently set forth.   400

DUKE

I am sorry that your leisure serves you not.

Antonio, gratify this gentleman,

For in my mind you are much bound to him.

Exeunt Duke and his train

BASSANIO *(to Portia)*

Most worthy gentleman, I and my friend

Have by your wisdom been this day acquitted   405

Of grievous penalties, in lieu whereof

Three thousand ducats, due unto the Jew,

We freely cope your courteous pains withal.

ANTONIO

And stand indebted over and above

In love and service to you evermore.   410

PORTIA

He is well paid that is well satisfied,

And I, delivering you, am satisfied,

And therein do account myself well paid.

My mind was never yet more mercenary.

I pray you, know me when we meet again.   415

I wish you well, and so I take my leave.

BASSANIO

Dear sir, of force I must attempt you further.

---

396 not] Q2; ~ to Q, F   *Exit Shylock*] ROWE; *Exit* Q, F   397 home with me] Q; with me home F   398 do] Q, F; *not in* Q2   403.1 *Exeunt*] Exit Q, F

398 **of** for, as at 2.5.37 (Abbott 174)
402 **gratify** remunerate, give a gratuity to (Onions)
403 **bound** (1) obliged, (2) indebted. The Duke's pun helps reduce the tension of the scene (NCS).
406 **in lieu whereof** in return for which
408 **cope** match (*OED cope v.*[2] 8), but possibly 'give in recompense' (*OED cope v.*[3] 2) (Merchant, NCS)

415 **know me** (1) recognize me, (2) consider this as an introduction, (3) (to Bassanio) know me carnally, i.e. sleep with me (this sense depends on the audience's awareness of the disguise)
417 **of force** perforce. Bassanio detains Portia, who has started to leave with Nerissa.
attempt urge

Take some remembrance of us as a tribute,
Not as fee. Grant me two things, I pray you:
Not to deny me, and to pardon me.   420

PORTIA

You press me far, and therefore I will yield.
⌈*To Antonio*⌉ Give me your gloves, I'll wear them for
    your sake.
(*To Bassanio*) And for your love, I'll take this ring from
    you.
Do not draw back your hand. I'll take no more,
And you in love shall not deny me this!   425

BASSANIO

This ring, good sir? Alas, it is a trifle.
I will not shame myself to give you this.

PORTIA

I will have nothing else but only this;
And now methinks I have a mind to it.

BASSANIO

There's more depends on this than on the value.   430
The dearest ring in Venice will I give you,
And find it out by proclamation.
Only for this, I pray you pardon me.

PORTIA

I see, sir, you are liberal in offers.
You taught me first to beg, and now methinks   435
You teach me how a beggar should be answered.

BASSANIO

Good sir, this ring was given me by my wife,
And when she put it on she made me vow
That I should neither sell nor give nor lose it.

---

419 fee] Q, F; a fee Q2   425 this!] ~? Q, F; ~. Q2   427 this.] Q2, F; ~? Q   429
it.] Q2, F; ~? Q   433 me.] F; ~? Q1–2

---

420 **pardon me** i.e. for being insistent
422.1, 423.1 Besides Clarendon's jus-
    tifications (cited by Furness) for the
    added stage directions, e.g. that the
    Duke has asked *Antonio* to 'gratify'
    Portia (402), note that Bassanio begs
    her to 'Take some remembrance of *us*'
    (418), meaning both Antonio and
    himself.
423 **for your love** for your sake (Schmidt;

cf. 1.3.167); but note the comic irony
    here (NS).
425 **love** kindness, but the irony con-
    tinues
430 **There's . . . value** i.e. more than the
    cost of the ring is at stake (Brown)
436 **how . . . answered** Compare 'A
    shameless beggar should have a shame-
    less denial' (Tilley A345; Dent, 51).

PORTIA

That 'scuse serves many men to save their gifts.

An if your wife be not a madwoman,

And know how well I have deserved this ring,

She would not hold out enemy for ever

For giving it to me. Well, peace be with you!

                      Exeunt Portia and Nerissa

ANTONIO

My Lord Bassanio, let him have the ring.   445

Let his deservings and my love withal

Be valued 'gainst your wife's commandement.

BASSANIO

Go, Graziano, run and overtake him.

Give him the ring, and bring him, if thou canst,

Unto Antonio's house. Away! Make haste!   450

                      Exit Graziano

Come, you and I will thither presently.

And in the morning early will we both

Fly toward Belmont. Come, Antonio.   *Exeunt*

4.2   *Enter Portia and Nerissa, still disguised*

PORTIA

Inquire the Jew's house out, give him this deed,

And let him sign it. We'll away tonight

And be a day before our husbands home.

This deed will be well welcome to Lorenzo.

      Enter Graziano

GRAZIANO

Fair sir, you are well o'erta'en.   5

My Lord Bassanio upon more advice

Hath sent you here this ring, and doth entreat

441 An] CAPELL; and Q, F   442 this] Q, F; the Q2   444.1 *Portia and Nerissa*] THEOBALD (*subs.*); *not in* Q, F   447 'gainst] Q; against F

    4.2] *after* Capell; *not in* Q, F   0.1 *Portia and*] F; *not in* Q   *still disguised*] RIVERSIDE (*subs.*); *not in* Q, F   4 Lorenzo.] Q2, F; ~? Q   5 o'erta'en] Q, F; overtaken MALONE

440 **'scuse** variant of *excuse*

443 **hold . . . ever** persist in being your enemy

447 **commandement** an old spelling and pronunciation, as in I *Henry VI* 1.4.20, preserved for the sake of metre

453 **Fly** hasten

4.2.1 **this deed** i.e. the deed of gift

(4.1.384–6, 390–93). Portia then puns on *deed* = 'act' (4).

5 **Fair . . . o'erta'en** a short verse line, allowing Graziano to catch his breath after running to catch them up (Brown, NCS)

6 **more advice** further consideration

Your company at dinner.

PORTIA   That cannot be.
His ring I do accept most thankfully,
And so I pray you tell him. Furthermore,   10
I pray you show my youth old Shylock's house.

GRAZIANO
That will I do.

NERISSA   Sir, I would speak with you.
(*Aside to Portia*) I'll see if I can get my husband's ring,
Which I did make him swear to keep forever.

PORTIA (*aside to Nerissa*)
Thou mayst, I warrant. We shall have old swearing   15
That they did give the rings away to men;
But we'll outface them, and outswear them too.—
Away! Make haste. Thou know'st where I will tarry.

                                                        Exit

NERISSA
Come, good sir, will you show me to this house?

                                                      Exeunt

5.1   *Enter Lorenzo and Jessica*

LORENZO
The moon shines bright. In such a night as this,
When the sweet wind did gently kiss the trees
And they did make no noise, in such a night
Troilus methinks mounted the Trojan walls
And sighed his soul toward the Grecian tents,   5
Where Cressid lay that night.

---

9 His] Q, F; This Q2   18.1 *Exit*] OXFORD; *not in* Q, F   19 *Exeunt*] F; *not in* Q
    5.1] *after* Rowe; *Actus Quintus* F; *not in* Q   1 The . . . this] Q, F; *two lines breaking
after* 'bright' Q2

---

15 **old** great, plentiful (Onions), a collo-
    quialism as in *Merry Wives* 1.4.5. On
    the significance of the ring plot, see
    Introduction, pp. 54–6.
17 **outface** stare down; hence, shame or
    silence, especially by boldness, assur-
    ance, or arrogance (*OED* 1); see 5.1.
    142–248.
5.1.1–14 **The moon . . . Aeson** The clas-
    sical allusions to Thisbe, Dido, and
    Medea derive ultimately from Ovid and
    Virgil, but those to Troilus are from

medieval sources. Possibly Chaucer
suggested them all to Shakespeare,
since Thisbe, Dido, and Medea appear
together in *The Legend of Good Women*,
which immediately follows *Troylus and
Creseyde* in old collections. The moonlight settings are specific to Chaucer's
treatment, although Medea's derives
from Golding's translation of Ovid, and
Dido's is transposed from Chaucer's
Ariadne (Hunter, cited by Furness; see
also Malone). The tales alluded to in-

JESSICA   In such a night
  Did Thisbe fearfully o'ertrip the dew
  And saw the lion's shadow ere himself
  And ran dismayed away.
LORENZO   In such a night
  Stood Dido with a willow in her hand   10
  Upon the wild sea banks and waft her love
  To come again to Carthage.
JESSICA   In such a night
  Medea gathered the enchanted herbs
  That did renew old Aeson.
LORENZO   In such a night
  Did Jessica steal from the wealthy Jew   15

11 wild sea banks] Q, F; wide Sea-banks ROWE; wild-sea ~ CAPELL   waft] Q, F; wav'd
THEOBALD

clude violations of a bond—to father, family, or city—at the command of love (John S. Coolidge, 'Law and Love in *The Merchant of Venice*', *SQ* 27 (1976), 261). In so far as the lovers (except Thisbe) are themselves betrayed, however, the allusions may seem inappropriate in the mouths of these newly-weds, even as they tease one another, joking half-nervously, perhaps, about love's transience (18–22; compare a related situation in *Dream* 1.1.173–6). For the dark undercurrents here and in other lyrical scenes in Shakespeare, see Merchant and J. L. Halio, 'Nightingales that Roar', in D. G. Allen and R. A. White (eds.), *Traditions and Innovations* (Newark, Del., 1990), 137–49. See also Wright, 139–40, for the metrical value of Lorenzo's and Jessica's 'shared' lines.

4–6 **Troilus . . . night** In *Troylus and Creseyde* 5.647–79 Chaucer describes Troylus looking at the moon, and facing toward the Greek tents where he believes his lover, Creseyde, sighs for *him*.

7–9 **Thisbe . . . away** Cf. Chaucer's *Legend of Good Women* 796–812, and Shakespeare's comic dramatization of the story in *Dream* 5.1.108–346.

7 **dew** This detail probably derives from an earlier passage in Chaucer's *Legend* (775) or from the 'dewie grasse' in Golding's translation of Ovid's *Metamorphoses* 4.102 (NS; J. D. Wilson,

SSur 10 (1957), 22).

8 **lion's shadow** a lioness in both Chaucer and Ovid, although Chaucer sometimes uses the masculine form (Brown). In *Dream* a lion frightens Thisbe away. The *shadow* may be cast by the moon, or it may refer to the lion's reflection in the fountain near where the lovers were to meet (Malone). **ere** before; the shadow is seen in advance of the lion himself

10–12 **Dido . . . Carthage** Dido was the queen of Carthage whom Aeneas loved and then abandoned (see Virgil's *Aeneid* and Ovid's *Heroides*, from which Chaucer drew his story in *Legend* 924–1367). Details here derive, however, from the story of Ariadne, *Legend* 2185–205 (Malone).

10 **willow** an emblem of forsaken love (see Desdemona's 'Willow Song', *Othello* 4.3.38–55). Ariadne did not carry a willow but a pole on which she stuck her kerchief (*Legend* 2202).

11 **wild** See Collation. Emendation is unnecessary; *wild* = 'without bounds, unconfined' (Sisson) or perhaps 'waste, desolate' (Brown).

waft beckoned, signalled (Onions)

13–14 **Medea . . . Aeson** Medea helped Jason win the golden fleece, then concocted a herbal brew that rejuvenated his father Aeson. Later, Jason abandoned her and their two children. Although the 'Legend of Hypsipyle and Medea' follows immediately after the 'Legend of Dido', this incident and its

And with an unthrift love did run from Venice
As far as Belmont.

JESSICA   In such a night
Did young Lorenzo swear he loved her well,
Stealing her soul with many vows of faith
And ne'er a true one.

LORENZO   In such a night   20
Did pretty Jessica, like a little shrew,
Slander her love, and he forgave it her.

JESSICA

I would outnight you, did nobody come.
But hark, I hear the footing of a man.

        Enter Stefano, a Messenger

LORENZO

Who comes so fast in silence of the night?   25

STEFANO A friend!

LORENZO

A friend—what friend? Your name, I pray you, friend?

STEFANO

Stefano is my name, and I bring word
My mistress will before the break of day
Be here at Belmont. She doth stray about   30
By holy crosses, where she kneels and prays
For happy wedlock hours.

LORENZO   Who comes with her?

STEFANO

None but a holy hermit and her maid.
I pray you, is my master yet returned?

24.1 *Stefano*] THEOBALD; *not in* Q, F   *a*] Q; *not in* F   26 STEFANO] VAR. 1803 (*subs.*);
Messen. Q; *Mes.* F   28, 33 STEFANO] VAR. 1803 (*subs.*); *Mess.* Q; *Messen.* Q2; *Mes.* F

language derive from Golding's translation of Ovid's *Metamorphoses* 7.162–
293, not Chaucer's *Legend*.
15 **steal** (1) sneak away, (2) rob; Jessica
    picks up the theme (19).

16 **unthrift** prodigal, spendthrift (Onions);
    cf. Sonnet 9.9
19 **Stealing** gaining possession of; cf.
    2.1.12 (Brown)
23 **outnight** out-do in mentioning nights
    (Onions)
24 **footing** steps, tread
24.1 As this messenger names himself,

editors since Theobald have followed
suit in the stage directions and speech
headings. He is probably the same one
with whom Portia bandied words
(2.9.84–97) (NCS).
25 **in silence** 'The article was frequently
    omitted before a noun already defined
    by another noun, especially in preposi-
    tional phrases' (Abbott 89); cf. 2.1.14,
    Sonnet 24.2.
31 **holy crosses** i.e. wayside shrines
33 **a holy hermit** He never appears, of
    course (see 88.1), and may simply be
    part of Portia's fiction.

LORENZO

He is not, nor we have not heard from him.   35

But go we in, I pray thee, Jessica,

And ceremoniously let us prepare

Some welcome for the mistress of the house.

    Enter Lancelot, the Clown

LANCELOT Sola, sola! Wo ha, ho! Sola, sola!

LORENZO Who calls?   40

LANCELOT Sola! Did you see Master Lorenzo? Master
Lorenzo, sola, sola!

LORENZO Leave hollering, man: here.

LANCELOT Sola! ⌈*Looking about*⌉ Where? Where?

LORENZO Here.   45

LANCELOT Tell him there's a post come from my master,
with his horn full of good news. My master will be here
ere morning.   *Exit*

LORENZO

Sweet soul, let's in, and there expect their coming.

38.1 *Lancelot, the*] BROWN (*after* Rowe); *not in* Q, F   39, 41, 44, 46 LANCELOT] ROWE;
Clowne Q, F   41 Master . . . Lorenzo,] CAMBRIDGE (*conj.* Thirlby); M. *Lorenzo,* & M.
Lorenzo Q, F; M. *Lorenzo,* M. *Lorenzo* Q2; M. *Lorenzo* and M. *Lorenza* F2; M. *Lorenzo* and
Mrs. *Lorenzo* F3–4; master Lorenzo and mistress Lorenza, POPE, CAPELL   43 hollering]
OXFORD *after* Q, F (hollowing); hallooing COLLIER   44 *Looking about*] This edition; *not
in* Q, F   45 Here.] Q2; ~? Q, F   48 morning.] ROWE; ~ sweete soule. Q, F, NCS   48
Exit] CAPELL (*subs.*); *not in* Q, F   49 Sweet soul, let's] VAR. 1785; Let's Q, F

---

37 **ceremoniously** transferred modifier, or
'adverbial hypallage' (Furness). Lorenzo means 'prepare a ceremonious welcome'.

39 **Sola . . . sola** Lancelot imitates a post-
horn (see 46 below); but compare
L.L.L. 4.1.148 (Brown). He is putting
on an act.

39 **Wo ha, ho** a falconer's call (*OED Wo* 1)

41–2 **Master . . . Master** See Collation. The
excrescent ampersand in Q may be a
misprint for a question mark (Furness),
unless here again Lancelot is being
pretentious. As the ensuing dialogue
indicates, he calls only to Lorenzo.

46 **post** courier

47 **horn . . . news** Lancelot compares his
'horn' to a cornucopia, or horn of
plenty (Kittredge).

49 **Sweet soul** See Collation. Some editors,
e.g. NCS, follow Q/F, but it is unlikely
that Lancelot would be addressing
Jessica as his speeches are clearly to

Lorenzo. Perhaps this part of the scene
was interpolated afterwards when Shakespeare realized an announcement of
Bassanio's arrival was necessary (NCS).
'Sweete soule' on the attached slip of
paper might have been indicated
where the insertion ended, but the
scribe or compositor could have mistaken these words as part of Lancelot's
speech and deleted the 'duplication' in
Lorenzo's (NS 106; Greg, *Editorial
Problem in Shakespeare* (Oxford, 1942),
123). Or the passage may have been
inserted 'to give the clown who played
Lancelot an opportunity of making the
theatre ring with his "sola!"' (NS
107). Or the passage was not an after-
thought, and the error originated
when a speaker's name was written
too low and 'sweete soule' got tacked
on to Lancelot's speech (Greg, *SFF*
258). See also 52 n.
expect await

And yet no matter. Why should we go in? 50
My friend Stefano, signify, I pray you,
Within the house, your mistress is at hand,
And bring your music forth into the air.   *Exit Stefano*
How sweet the moonlight sleeps upon this bank!
Here will we sit and let the sounds of music 55
Creep in our ears. Soft stillness and the night
Become the touches of sweet harmony.
Sit, Jessica. Look how the floor of heaven
Is thick inlaid with patens of bright gold.
There's not the smallest orb which thou behold'st 60
But in his motion like an angel sings,
Still quiring to the young-eyed cherubins.

51 Stefano] Q2; *Stephen* Q, F   I] Q; *not in* F   53 *Exit Stefano*] JOHNSON; *not in* Q, F
59 patens] Q, F; pattents Q2; patterns F2; patines MALONE

52 **your mistress** Since Lorenzo does not mention the master of the house, the announcement of Bassanio's return (38.1–48) is 'indeed an interpolation' (NCS). But has Bassanio had time enough to establish himself as master of Belmont?

53 **music** i.e. the musicians

57 **Become** befit, suit
touches strains, notes, produced by fingering a musical instrument (Onions); see 67 n.

58 **floor of heaven** i.e. the sky. The canopy over the stage, or 'heavens', of an Elizabethan theatre was painted with stars and heavenly signs (Merchant).

59 **patens** small dishes, often made of gold, from which the eucharist is served (Malone). They suggest tiles, but as circles they are also an image of harmony (Danson, 19).

60–5 **There's . . . hear it** The singing of the stars, or planets, in their fixed spheres was a commonplace, originating in ancient thought, e.g. Plato's *Republic* and Plutarch's *De re musica*. While the human soul is immured in the corporeal body ('this muddy vesture of decay') it is unable to hear or respond to celestial music, as the cherubim do. Compare Montaigne, *Of Custom*, bk. 1, ch. 22 (Florio's translation, quoted by NS): 'what philosophers deem of the celestial music, which is that the bodies of its circles, being solid smooth, and in their rolling motion touching and rubbing one against another, must of necessity produce a wonderful harmony. . . . But that universally the hearing senses of these low world's creatures, dizzied and lulled asleep . . . by the continuation of that sound, how loud and great soever it be, cannot sensibly perceive or distinguish the same.' In 'Realization', 208, Brown says: 'Shakespeare reminds the audience that behind an earthly harmony lies another that is perfect, that behind the conclusions of the fifth act is a further peace, one which may resolve and contain all apparent discords.' Accordingly, it is important that a 'reverently harmonious music' and not anything 'perky or sugary' should be played here.

61 **his** its

62 **still quiring** always singing together in consort. Compare Te Deum Laudamus in the Book of Common Prayer, sung or said at Morning Prayer: 'To thee all angels cry aloud: the heavens and all the powers therein. To thee cherubin and seraphin continually do cry' (Noble).

young-eyed cherubins i.e. with sight ever young. Medieval tradition endowed cherubim with powerful sight (Verity, cited by NS). Cf. *Hamlet* 4.3.50: 'I see a cherub sees them', and Ezek. 10, which uses the common, if irregular, plural 'Cherubims' and mentions their eyes (12).

Such harmony is in immortal souls,
But whilst this muddy vesture of decay
Doth grossly close it in, we cannot hear it.   65

　　⌈Enter Musicians⌉

(*To the Musicians*) Come, ho! and wake Diana with a
　　hymn.
With sweetest touches pierce your mistress' ear
And draw her home with music.

　　　Music plays

JESSICA

I am never merry when I hear sweet music.

LORENZO

The reason is your spirits are attentive,   70
For do but note a wild and wanton herd
Or race of youthful and unhandled colts
Fetching mad bounds, bellowing and neighing loud,
Which is the hot condition of their blood;
If they but hear perchance a trumpet sound,   75
Or any air of music touch their ears,
You shall perceive them make a mutual stand,

---

65 it in] Q; in it Q2, F; us in ROWE 1714   65.1] MALONE (*after* Capell]; *not in* Q, F
66 *To the Musicians*] OXFORD; *not in* Q, F   68.1] Q2; *play Musicke* Q; (*after l.* 69) F

---

63 **harmony** (1) concord, (2) power to
appreciate music (Brown)

65 **it . . . it** Both pronouns may take
'harmony' (63) as their antecedent,
although the first 'it' may refer collectively to 'immortal souls' (Brown, NCS).

66 **wake** Either (1) the moon has gone
behind a cloud (see 92; NCS), or (2)
wake = 'keep her vigil' (NS), or (3)
wake = 'excite' (*Much Ado* 5.1.104:
'we will not wake your patience').
　　Diana goddess of the moon, but also
of chastity

67 **touches** i.e. fingering of your instru-
ments

68 **draw . . . music** See 80 n.

69 **I . . . music** Jessica's last line in the
play has encouraged some modern directors to interpret her role and her
reaction to events in a negative or
emotionally disturbed fashion, especially in this scene. (See Introduction,
p. 74.) But Jessica means that music
puts her into a contemplative or reflective attitude, and Lorenzo's reply explains why. Compare Danson, 187–8,

who cites the effect of 'human music'
described in Boethius, *De institutione
musica*.

70 **spirits** faculties of perception or reflection, as in *Othello* 3.4.62 (*OED* 18),
but also suggesting the mind's procliv-
ity to depression or exaltation (*OED*
17)
　　attentive observant, heedful (*OED*).
When joyful, the mental faculties are
lively; when observant, they are tran-
quil. Cf. 86–7, below.

71–9 **note . . . music** The image of the
colts stopped in their antics by the
sound of music illustrates Lorenzo's
point: music makes even beasts seem
quietly reflective. Cf. *Tempest* 4.1.175–
8: 'Then I beat my tabor, | At which
like unbacked colts they pricked their
ears, | Advanced their eyelids, lifted up
their noses | As they smelt music' (Ma-
lone).

72 **race** herd or stud (Onions)
　　unhandled untamed, not broken in

73 **Fetching** performing (Onions)

77 **mutual** common

Their savage eyes turned to a modest gaze
By the sweet power of music. Therefore the poet
Did feign that Orpheus drew trees, stones, and floods,   80
Since naught so stockish, hard, and full of rage,
But music for the time doth change his nature.
The man that hath no music in himself,
Nor is not moved with concord of sweet sounds,
Is fit for treasons, stratagems, and spoils;   85
The motions of his spirit are dull as night,
And his affections dark as Erebus.
Let no such man be trusted. Mark the music.
    Enter Portia and Nerissa, as themselves, approaching
PORTIA

That light we see is burning in my hall.
How far that little candle throws his beams—   90
So shines a good deed in a naughty world.
NERISSA

When the moon shone, we did not see the candle.
PORTIA

So doth the greater glory dim the less.
A substitute shines brightly as a king
Until a king be by, and then his state   95

82 the] Q; *not in* F   87 Erebus] F2; *Terebus* Q; *Erobus* F   88.1] *as themselves*] OXFORD;
not in Q, F   *approaching*] This edition; *at a distance* JOHNSON; *not in* Q, F   92 candle.]
Q2; ~? Q, F

79 **the poet** Ovid
80 **feign** invent (a story) (*OED* 2)
    Orpheus drew trees The story of the
    power of Orpheus' music to attract
    trees, etc. appears in *Metamorphoses*
    10.88–111.
81 **stockish** blockish, unfeeling (Onions)
83–5 **The man . . . spoils** Compare Shy-
    lock, 2.5.28–36, and Cassius in *Caesar*
    1.2.205, neither of whom loves music.
    The concept derives from Neoplatonism.
85 **stratagems** deeds of great violence, as
    in *Romeo* 3.5.209 (Onions)
    spoils rapine, plunder; compare 'spoils
    of war'
86 **motions . . . night** 'The impulses of his
    mind are as sinisterly impenetrable as
    is darkness' (NCS). This may also be a
    cue for the moon's becoming obscured
    (see 89–92).
87 **affections** thoughts, feelings, as at
    1.1.16

Erebus region of darkness in the
underworld
91 **So . . . world** Cf. Matt. 5: 16: 'Let
your light so shine before men, that
they may see your good works' (Halliwell, cited by Furness).
92–3 **When . . . less** Cf. Tilley S826.2:
'To be like stars to the moon' (Dent,
218). The saying, which may have
originated in Horace's *Odes* 1, 12, 47,
refers to Augustus as 'the moon
among lesser lights'. As in 80, the Ode
refers to Orpheus drawing trees and
streams by his music (NCS). The moon
has apparently become obscured; cf.
109–10 n.
95–7 **his state . . . waters** The image shifts
from light to water; the substitute's
'state', or dignity, disappears like a
stream's water flowing into the ocean.
Cf. 'All rivers run into the sea' (Tilley
R140; Dent, 202).

Empties itself as doth an inland brook
Into the main of waters. Music, hark!

NERISSA

It is your music, madam, of the house.

PORTIA

Nothing is good, I see, without respect.
Methinks it sounds much sweeter than by day.   100

NERISSA

Silence bestows that virtue on it, madam.

PORTIA

The crow doth sing as sweetly as the lark
When neither is attended, and I think
The nightingale, if she should sing by day
When every goose is cackling, would be thought   105
No better a musician than the wren.
How many things by season seasoned are
To their right praise and true perfection!
⌈*She sees Lorenzo and Jessica*⌉
Peace, ho! ⌈*Music ceases*⌉ The moon sleeps with Endy-
  mion
And would not be awaked.

LORENZO   That is the voice,   110
Or I am much deceived, of Portia.

97 hark!] ~. Q; ~. *Musicke. (as SD)* F   98 house.] Q2, F; ~? Q   100 day.] Q2; ~? Q,
F   101 madam.] Q2, F; ~? Q   106 wren.] Q2; ~? Q (Renne), F   108.1] OXFORD
(*after* Capell); *not in* Q, F   109 Peace, ho!] MALONE; ~, how∧ Q, F; ~! How POPE
Music ceases] OXFORD; *after* 'awak'd', *l.* 110 F; *not in* Q

97 **main of waters** ocean
99 **respect** relationship, reference (Onions).
  Cf. *Hamlet* 2.2.250: 'there's nothing
  good or bad but thinking makes it so.'
  The present circumstances make the
  music seem sweeter to Portia, as she
  and Nerissa continue discussing the
  theme of relationship, or relative ap-
  preciation.
101 **virtue** power, efficacy; cf. 199, below
103 **attended** noticed (Schmidt), or accompanied by others (Brown)
107 **How . . . seasoned** So many things
  are qualified (*OED seasoned ppl.a.* 3b) by
  their circumstances (*OED season sb.* 12).
108 **true perfection** 'Seasoned perfection
  is a paradox, impenetrable by natural
  reason . . . [A person] may find true,
  if "seasoned", perfection in the concord of human terms' (Oz, 105).

109 **Peace, ho!** See Collation. As F's stage
  direction indicates, Portia apparently
  commands the musicians to stop
  playing. Malone compared similar
  commands in *Romeo* 4.4.92, *As You
  Like It* 5.4.123, and *Measure* 1.4.6,
  3.1.44. *Ho* was frequently printed
  'how' or 'howe', as at 2.6.25.
109–10 **The moon . . . awaked** Endymion
  was a young shepherd beloved of
  Selene (Diana), the moon goddess,
  who put him asleep forever in a cave
  so she could visit him when she
  pleased. Capell's stage direction indicates that Portia is prompted by seeing
  Lorenzo and Jessica asleep in each
  other's arms. Portia's lines and Lorenzo's response also indicate that the
  moon is obscured (see 92–3 n.).

PORTIA

He knows me as the blind man knows the cuckoo—
By the bad voice.

LORENZO   Dear lady, welcome home.

PORTIA

We have been praying for our husbands' welfare,
Which speed, we hope, the better for our words.   115
Are they returned?

LORENZO   Madam, they are not yet,
But there is come a messenger before
To signify their coming.

PORTIA   Go in, Nerissa,
Give order to my servants that they take
No note at all of our being absent hence;   120
Nor you, Lorenzo; Jessica, nor you.
        ⌈*A tucket sounds*⌉

LORENZO

Your husband is at hand. I hear his trumpet.
We are no tell-tales, madam. Fear you not.

PORTIA

This night methinks is but the daylight sick.
It looks a little paler. 'Tis a day   125
Such as the day is when the sun is hid.

        Enter Bassanio, Antonio, Graziano, and their fol-
        lowers. Graziano and Nerissa ⌈*embrace and*⌉ *speak*
        apart to one another.

112–13 He . . . voice] Q; *lines break* 'knowes | The' Q2; *as prose* F   113 voice.] Q2;
~? Q, F   home.] Q2; ~? Q, F   114 husbands' welfare] Q, F; husband health Q2   118
coming.] Q2, F; ~? Q   121 A tucket sounds] F; *not in* Q   126.2 Graziano and . . .
another] OXFORD (*subs.*); *not in* Q, F

---

114 **praying** See 30–2, above.
115 **speed** prosper
118–20 **Go in . . . hence** Nerissa does not
exit, as the tucket interrupts; in any
case, the household have already been
partially apprised of the plot (3.4.37),
which is fully revealed before Bassanio
meets any of Portia's servants. Or Nerissa goes in (119), Graziano looks for
her inside, and they re-emerge (142)
quarrelling about the ring (NCS).
121.1 **tucket** flourish on a trumpet, from
Italian *toccata* (Steevens)

122 **his trumpet** Tuckets were individ-
ualized; compare *Lear* 2.2.355–6. But
Bassanio is the only private citizen in
Shakespeare to have one (NCS).
124–6 **This . . . hid** Portia's small talk
suggests that the moon has reap-
peared; see 142.
126.1–3 Bassanio's entrance with the
others should come after 123 or at
least during Portia's speech if his lines
(127–8) are, as intended, a response
to hers.

BASSANIO

We should hold day with the Antipodes,

If you would walk in absence of the sun.

PORTIA

Let me give light, but let me not be light;

For a light wife doth make a heavy husband,   130

And never be Bassanio so for me.

But God sort all! You are welcome home, my lord.

BASSANIO

I thank you, madam. Give welcome to my friend.

This is the man, this is Antonio,

To whom I am so infinitely bound.   135

PORTIA

You should in all sense be much bound to him,

For as I hear he was much bound for you.

ANTONIO

No more than I am well acquitted of.

PORTIA

Sir, you are very welcome to our house.

It must appear in other ways than words,

Therefore I scant this breathing courtesy.   140

GRAZIANO (*to Nerissa*)

By yonder moon I swear you do me wrong.

In faith, I gave it to the judge's clerk.

Would he were gelt that had it, for my part,

Since you do take it, love, so much at heart.   145

PORTIA

A quarrel, ho, already! What's the matter?

132 You are] Q, F; y'are Q2   142 *to Nerissa*] ROWE (*after l. 143*); *not in* Q, F

127–8 We . . . sun Bassanio's courtly
   compliment eases him into conversation with his wife. Malone paraphrases: 'If you would always walk in
   the night it would be day with us, as
   it now is on the other side of the globe'
   (the Antipodes).
129 give light . . . be light Shakespeare
   and his contemporaries loved to pun
   on *light* = wanton.
130 heavy unhappy. Compare 'A good
   wife makes a good husband' (Tilley
   W351; Dent, 249).
132 God sort all let it be as God wishes.
   Portia anticipates Bassanio's discomfit-

ure over the ring (177–248).
135–7 bound . . . bound . . . bound (1)
   tied (in bonds of friendship), (2) indebted, (3) pledged, (4) imprisoned
136 in all sense (1) in all reason, or (2)
   in every way
138 acquitted of requited (*OED* 6, as in
   Henry V 2.2.141)
141 breathing verbal
142 By yonder moon an inapt oath, since
   the moon was an emblem of incon-
   stancy, as Juliet notes (*Romeo* 2.1.
   151–3)
144 gelt gelded

GRAZIANO

    About a hoop of gold, a paltry ring
    That she did give me, whose posy was
    For all the world like cutler's poetry
    Upon a knife—'Love me, and leave me not'.   150

NERISSA

    What talk you of the posy or the value?
    You swore to me when I did give it you
    That you would wear it till your hour of death,
    And that it should lie with you in your grave.
    Though not for me, yet for your vehement oaths   155
    You should have been respective and have kept it.
    Gave it a judge's clerk! No, God's my judge,
    The clerk will ne'er wear hair on's face that had it.

GRAZIANO

    He will, an if he live to be a man.

NERISSA

    Ay, if a woman live to be a man.   160

GRAZIANO

    Now, by this hand, I gave it to a youth,
    A kind of boy, a little scrubbèd boy,
    No higher than thyself, the judge's clerk,
    A prating boy, that begged it as a fee.
    I could not for my heart deny it him.   165

PORTIA

    You were to blame, I must be plain with you,
    To part so slightly with your wife's first gift,
    A thing stuck on with oaths upon your finger
    And so riveted with faith unto your flesh.
    I gave my love a ring and made him swear   170
    Never to part with it; and here he stands.
    I dare be sworn for him he would not leave it,
    Nor pluck it from his finger for the wealth

---

148 posy] Q; poesie Q2, F   152 give it] Q2, F; giue Q   153 your] Q; the F   157 clerk!] NS; ~? OXFORD; Clarke: Q, F   No . . . judge] Q; but well I know F

148 **posy** 'A short motto, originally a line of verse . . . inscribed on a knife, within a ring . . . etc.' (*OED* I)
149 **cutler's poetry** i.e. poor stuff
150 **leave** part with, as at 172 and *Two*
Gentlemen 4.4.72, where a similar situation obtains (NCS)
155 **Though** if
156 **respective** careful, regardful
162 **scrubbèd** stunted

That the world masters. Now, in faith, Graziano,
You give your wife too unkind a cause of grief.   175
An 'twere to me, I should be mad at it.
BASSANIO (*aside*)
    Why, I were best to cut my left hand off
    And swear I lost the ring defending it.
GRAZIANO
    My Lord Bassanio gave his ring away
    Unto the judge that begged it, and indeed   180
    Deserved it, too; and then the boy, his clerk,
    That took some pains in writing, he begged mine,
    And neither man nor master would take aught
    But the two rings.
PORTIA (*to Bassanio*) What ring gave you, my lord?
    Not that, I hope, which you received of me.   185
BASSANIO
    If I could add a lie unto a fault,
    I would deny it; but you see my finger
    Hath not the ring upon it. It is gone.
PORTIA
    Even so void is your false heart of truth.
    By heaven, I will ne'er come in your bed   190
    Until I see the ring.
NERISSA (*to Graziano*) Nor I in yours
    Till I again see mine.
BASSANIO   Sweet Portia,
    If you did know to whom I gave the ring,
    If you did know for whom I gave the ring,
    And would conceive for what I gave the ring,   195
    And how unwillingly I left the ring
    When naught would be accepted but the ring,

---

177 *aside*] THEOBALD; *not in* Q, F   191 ring.] Q2, F; ~? Q   *to Graziano*] OXFORD; *not in*
Q, F   191–2 Nor . . . mine] Q; *one line* F   192 mine.] Q2, F; ~? Q

174 **masters** possesses
176 **An** If
    mad frantic, beside myself
193–7, 199–202 **If . . . ring, If . . . ring**
    Bassanio uses rhetorical figures of
    repetition, including anaphora (begin-
    ning a series of clauses with the same
    word) and epistrophe (ending with the

same), which Portia follows—and betters him in the argument. The same
figures are used, though to different
effect, in *Richard III* 4.4.40–6. Here,
they enhance the comedy of the situation.
196 **left** parted with; see 150 n. and 172

You would abate the strength of your displeasure.
PORTIA
   If you had known the virtue of the ring,
   Or half her worthiness that gave the ring,   200
   Or your own honour to contain the ring,
   You would not then have parted with the ring.
   What man is there so much unreasonable,
   If you had pleased to have defended it
   With any terms of zeal, wanted the modesty   205
   To urge the thing held as a ceremony?
   Nerissa teaches me what to believe:
   I'll die for't, but some woman had the ring!
BASSANIO
   No, by my honour, madam, by my soul,
   No woman had it, but a civil doctor,   210
   Which did refuse three thousand ducats of me
   And begged the ring; the which I did deny him
   And suffered him to go displeased away—
   Even he that had help up the very life
   Of my dear friend. What should I say, sweet lady?   215
   I was enforced to send it after him.
   I was beset with shame and courtesy.
   My honour would not let ingratitude
   So much besmear it. Pardon me, good lady,
   For by these blessèd candles of the night,   220
   Had you been there, I think you would have begged
   The ring of me to give the worthy doctor.
PORTIA
   Let not that doctor e'er come near my house.

198 displeasure.] Q2; ~? Q, F   208 ring!] ~? Q, F; ~. Q2   209 my honour] Q; mine honour F   213 displeased away] Q, F; away displeasd Q2   214 had held up] Q, F; did uphold Q2   220 For] Q; And F 222 doctor.] Q2; ~? Q, F

199 **virtue** power; cf. 101, above
201 **contain** retain; cf. 4.1.49
205 **modesty** considerateness, freedom from arrogance; cf. *Twelfth Night* 2.1.10–12: 'I perceive in you so excellent a touch of modesty that you will not extort from me what I am willing to keep in.'
206 **held as a ceremony** regarded as an important symbol (cf. *Henry V* 4.1.104: '[the king's] ceremonies laid by . . .')

210 **civil doctor** doctor of civil law
211 **Which** who
214 **held up** sustained, preserved, as in *Dream* 3.2.240
216–9 **I . . . besmear it** 'It indicates the gentleman and the soldier in Bassanio, that he does not expose Antonio as the one that "enforced" him to give the ring' (Allen, cited by Furness).
220 **candles** stars, as in *Romeo* 3.5.9

Since he hath got the jewel that I loved,
And that which you did swear to keep for me,   225
I will become as liberal as you.
I'll not deny him anything I have,
No, not my body nor my husband's bed.
Know him I shall, I am well sure of it.
Lie not a night from home; watch me like Argus.   230
If you do not, if I be left alone,
Now, by mine honour, which is yet mine own,
I'll have that doctor for my bedfellow.

NERISSA
And I his clerk. Therefore be well advised
How you do leave me to mine own protection.   235

GRAZIANO
Well, do you so. Let not me take him, then.
For if I do, I'll mar the young clerk's pen.

ANTONIO
I am th'unhappy subject of these quarrels.

PORTIA
Sir, grieve not you. You are welcome notwithstanding.

BASSANIO
Portia, forgive me this enforcèd wrong,   240
And in the hearing of these many friends,
I swear to thee, even by thine own fair eyes,
Wherein I see myself—

PORTIA   Mark you but that!
In both my eyes he doubly sees himself,
In each eye one. Swear by your double self,   245
And there's an oath of credit.

233 that] Q; the F   my] Q2, F; mine Q   239 Sir . . . notwithstanding] Q; *two lines breaking* 'you, | You' F   243 myself—] ROWE; my selfe. Q, F; my selfe.—F2   that!] ~? Q, F; ~. Q2

226 **liberal** (1) generous, (2) sexually free (Riverside)
229 **Know** (1) recognize, (2) know carnally, recalling 4.1.415
230 **Argus** Argus Panoptes (i.e. all-seeing), the hundred-eyed giant set to guard Io, daughter of the river god Inachus (Ovid, *Metamorphoses* 1.622–77).
232 **mine own** i.e. not surrendered to anyone, intact

236 **take** apprehend
237 **pen** (1) quill, (2) penis
240–3 **Portia . . . myself** Bassanio reverts to the courtly idiom with which he began (127–8), but Portia cuts him off, even more playfully responding to him than she did then.
245 **double** (1) twofold, (2) deceitful
246 **oath of credit** believable vow; Portia is being sarcastic

BASSANIO   Nay, but hear me.
Pardon this fault, and by my soul I swear
I never more will break an oath with thee.

ANTONIO
I once did lend my body for his wealth
Which, but for him that had your husband's ring,   250
Had quite miscarried. I dare be bound again,
My soul upon the forfeit, that your lord
Will never more break faith advisedly.

PORTIA
Then you shall be his surety. Give him this
And bid him keep it better than the other.   255

ANTONIO
Here, Lord Bassanio; swear to keep this ring.

BASSANIO
By heaven, it is the same I gave the doctor!

PORTIA
I had it of him. Pardon me, Bassanio,
For by this ring the doctor lay with me.

NERISSA
And pardon me, my gentle Graziano;
For that same scrubbèd boy, the doctor's clerk,   260
In lieu of this last night did lie with me.

GRAZIANO
Why, this is like the mending of highways
In summer where the ways are fair enough!
What, are we cuckolds ere we have deserved it?   265

PORTIA
Speak not so grossly. You are all amazed.
Here is a letter. Read it at your leisure.

249 his] Q; thy F   258 me] Q; *not in* F   264 where] Q, F; when COLLIER 1853
enough!] ~? Q; ~. Q2; ~: F

249 **wealth** (1) welfare, (2) material
  benefit
252 **my . . . forfeit** Having once pledged
  his body, Antonio now pledges his
  soul, something far more precious.
253 **advisedly** deliberately, intentionally
  (*OED* 4)
254-5 **Then . . . other** Sigurd Burckhardt,
  Shakespearean Meanings (Princeton, NJ,
  1968), 234-6, shows how the ring is
  the bond transformed into 'the gentle

bond'.
262 **In lieu of** in return for
263-4 **mending . . . enough** i.e. a work of
  supererogation. Graziano implies that
  the attempt to teach them a lesson was
  unnecessary; they have had no
  chance yet to be unfaithful.
265 **cuckolds** betrayed husbands
266 **grossly** (1) indelicately, (2) stupidly
  (Brown)
  amazed utterly confused, bewildered

225

It comes from Padua, from Bellario.
There you shall find that Portia was the doctor,
Nerissa there her clerk. Lorenzo here   270
Shall witness I set forth as soon as you
And even but now returned; I have not yet
Entered my house. Antonio, you are welcome,
And I have better news in store for you
Than you expect. Unseal this letter soon.   275
There you shall find three of your argosies
Are richly come to harbour suddenly.
You shall not know by what strange accident
I chancèd on this letter.

ANTONIO (*reads the letter*)   I am dumb!

BASSANIO
  Were you the doctor and I knew you not?   280

GRAZIANO
  Were you the clerk that is to make me cuckold?

NERISSA
  Ay, but the clerk that never means to do it,
  Unless he live until he be a man.

BASSANIO
  Sweet doctor, you shall be my bedfellow.
  When I am absent, then lie with my wife.   285

ANTONIO
  Sweet lady, you have given me life and living,
  For here I read for certain that my ships
  Are safely come to road.

PORTIA   How now, Lorenzo!
  My clerk hath some good comforts too for you.

272 even but] Q; but eu'n F   279 *reads the letter*] This edition; *not in* Q, F   dumb!]
~? Q; ~. Q2, F   288 Lorenzo!] ~? Q, F; ~, Q2

278–9 **You . . . letter** 'This beautiful
example of Shakespeare's dramatic im-
pudence has been severely criticized by
some pundits [e.g. Eccles, quoted by
Furness]. A more painstaking dramat-
ist, no doubt, would have brought on
the letter by a soaling "post" . . . and
thus ruined entirely the stillness and
intimacy of this wonderful finale' (NS).
Similarly, Shakespeare does not explain how Portia knows the contents

of the sealed letter.
279 **dumb** i.e. struck dumb with aston-
ishment (said as Antonio reads the
letter)
286 **Sweet . . . living** Portia has saved his
life and now his means of living; the
half of Shylock's fortune was Antonio's only 'in use' for Lorenzo and
Jessica (4.1.379–81).
288 **road** anchorage

NERISSA

Ay, and I'll give them him without a fee.

There do I give to you and Jessica   290

From the rich Jew a special deed of gift,

After his death, of all he dies possessed of.

LORENZO

Fair ladies, you drop manna in the way

Of starvèd people.

PORTIA   It is almost morning,

And yet I am sure you are not satisfied   295

Of these events at full. Let us go in,

And charge us there upon inter'gatories,

And we will answer all things faithfully.

GRAZIANO

Let it be so. The first inter'gatory   300

That my Nerissa shall be sworn on is

Whether till the next night she had rather stay

Or go to bed now, being two hours to day.

But were the day come, I should wish it dark

Till I were couching with the doctor's clerk.   305

Well, while I live I'll fear no other thing

So sore as keeping safe Nerissa's ring.   *Exeunt*

296 I am] Q, F; Ime Q2   297 Let us] Q, F; Let's Q2   298 inter'gatories] F; intergotories
Q   300 inter'gatory] F; intergory Q *uncorr.*; intergotory Q *corr.*   305 Till] Q, F; That
Q2   doctor's] Q, F; *not in* Q2

292 **deed of gift** See 4.1.384–6.
294–5 **you drop . . . people** Lorenzo and
 Jessica, a spendthrift couple, have apparently exhausted their funds—or
 soon will. See 15–17 above, 3.1.101–
 105, 4.1.379 n.
294 **manna** food miraculously provided
 for the Israelites during the Exodus.
 See Exod. 16: 15, where the marginal
 gloss indicates the food is 'a gift' (Merchant).
296–7 **satisfied . . . full** 'so fully informed
 about all these occurrences as to feel
 no further curiosity' (Kittredge)
298 **charge . . . inter'gatories** Portia reverts to legal jargon. Witnesses were
 compelled (*charged*) upon oath (301) to
 answer specific questions (*interrogatories*), then as now.
307 **ring** (1) piece of jewellery, (2) vulva.
 The romantic comedy appropriately

ends on another bit of bawdy punning,
an allusion to the old story of Hans
Castorp's ring (see Leslie Fiedler, *The
Stranger in Shakespeare* (New York,
1972), 136, and cf. Harry Levin, 'A
Garden in Belmont', in W. R. Elton
and William B. Long (eds.), *Shakespeare
and Dramatic Tradition* (Newark, Del.,
1989), 30). Granville-Barker found a
practical reason for giving Graziano
the last speech (against tradition): the
others 'must pace off the stage in their
stately Venetian way, while Gratiano's
harmless ribaldry is tossed to the audience as an epilogue. Then he and Nerissa, now with less dignity than ever
to lose, skip after' (p. 107). But the
ending is often staged quite differently
(see Leggatt, 149, and stage history,
above).

$\frac{45}{50}$ The notes I could read are excellent! Baxter, despite my great esteem for you, if an assignment is this late again, marks will have to be deducted!

SPEECH PREFIXES FOR SHYLOCK

(by line number, edition, signature and compositor)

Line no.	Q1	sig.	Comp.	Q2	sig.	Comp.	F	sig.	Comp.
1.3.1	Shy	B2	Y	Shy	B2	H	Shy.	O5b	D
.3	Shy.				B2^v	H			
.6	Shy.								
.9	Shy.								
.12	Shy.								
.15	Shylocke								
.28	Iew.	B2^v	Y				Iew.	O5^{v}a	D
.31	Iew.								
.38	Iew.								
.50	Shyl.				B3	H	Shy.		
.62	Shy.	B3	Y						
.64	Shyl.								
.68	Shy.								
.73	Shyl.				B3^v	H			
.93	Shyl.	B3^v	Y					O5^{v}b	D
.100	Shy.								
.103	Shyl.								
.133	Shy.	B4	Y		B4	H			
.140	Shyl.								
.157	Shy.				B4^v	H		O6a	D
.169	Shy.	B4^v	Y						
2.5.1	Iewe.	C4^v	X		D1	H	Iew.	P1b	D
.7	Shy.	D1	Y		D1^v	H	Shy.		
.11	Shy.								
.21	Shy.								
.28	Shy.							P1^{v}a	C
.43	Shyl.	D1^v	Y						
.45	Shy.				D2	H			
3.1.23	Shy.	E2	X		E3	G	Shy.	P3a	D
.30	Shy.								
.32	Shy.								
.35	Shy.								
.41	Shy.	E2^v	X						
.50	Shyl.			Shyl.					
.75	Shy.			Shy.	E3^v	G			
.79	Shylocke.	E3	X						
.94	Shy.				E4	G		P3b	D
.96	Shy.								
.99	Shy.								
.103	Shy.								
.109	Shy.								

Line no.	QI	sig.	Comp.	Q2	sig.	Comp.	F	sig.	Comp.
.113	Shy.								
.118	Shy.								
3.3.1	Iew.	F4	Y	Iew.	GI	G	Iew.	P4^{v}a	C
.4	Iew.							P4^{v}b	C
.12	Iew.								
4.1.33	Iewe.	G3^v	X	Iew.	G4^v	G	Iew.	P5^{v}b	B
.64	Iewe.			Shy.	HI	H			
.66	Iewe.								
.68	Iewe.								
.84	Iewe.	G4	X	Iew.	HI^v	H			
.88	Iewe.								
.121	Iewe.	G4^v	X		H2	H		P6a	B
.126	Iewe.								
.138	Iewe.								
.173	Iew.	HI	Y		H2^v	H		P6b	B
.180	Shy.	HI^v	Y	Shy.	H3				
.203	Shy.						Shy.		
.220	Shy.	H2	Y		H3^v	H	Iew.	P6^{v}a	B
.223	Shy.								
.225	Shy.						Shy.		
.232	Shy.						Iew.		
.243	Shy.								
.247	Iew.	H2^v	Y		H4	H			
.249	Iew.								
.253	Iew.								
.256	Iew.								
.259	Iew.								
.292	Iew.	H3	Y	Iew.	H4^v	H		P6^{V}b	C
.298	Iew.								
.301	Iew.								
.311	Shy.			Shy.			Shy.		
.315	Iew.			Iew.			Iew.		
.332	Shy.	H3^v	Y	Shy.	I1	G	Shy.		
.338	Shy.								
.341	Shy.								
.370	Shy.	H4	Y		I1^v	G		QIa	C
.390	Shy.	H4^v	Y		I2	G			
.391	Shy.								

INDEX

THIS index lists points of more than routine interest but does not duplicate headings or sub-headings in the Introduction, nor does it list editorial decisions in Editorial Procedures. Biblical and proverbial phrases and allusions, respectively, are grouped. So are discussions of puns, quibbles, and other ambiguous words; rhyme and verse forms; and legal matters. Citations from Shakespeare's other works are not listed, nor are characters' names. For words that appear several times in the text with the same meaning, only the first use is listed, but multiple meanings of the same word are listed accordingly. While references to the Commentary are primarily to the lemmata, other significant discussions are also included from them and from the General Introduction under key words, such as 'melancholy'. Asterisks identify words which supplement the *Oxford English Dictionary*.

a (= he), 2.1.30
a many, 3.5.63
abode, 2.6.21
Abram, 1.3.69
abroad, 3.3.10
accomplishèd, 3.4.61
accordingly, 2.1.0.3
account, 3.2.155, 157
accoutred, 3.4.63
acquitted of, 5.1.138
addressed me, 2.9.18
Admiral's Men, p. 28
advantage, 1.3.67
adventuring, 1.1.143
advisedly, 5.1.253
Aeson, 5.1.14
affection wondrous sensible, 2.8.48
affections, 1.1.16; 3.1.57; 4.1.49
agitation, 3.5.4
ague, 1.1.23
alabaster, 1.1.84
Alcides, p. 36; 2.1.35
Aldridge, Ira, p. 80
Alexander, Bill, pp. 77–9
Alfi, Yossi, p. 83
all yours, 3.2.18
aloof, 3.2.42
amazed, 5.1.266
ambitious head, 2.7.44

amity, 3.2.30; 3.4.3
an, 1.2.45; an't, 2.2.55
ancient Roman honour, 3.2.293
Andrew, p. 27; 1.1.27
angel, 2.7.56
answer, 4.1.2
apparent, 4.1.20
apple, 1.3.98
argosies, 1.1.9
Argus, 5.1.230
Aristotle, *Politics*, 1.3.130
armed, 4.1.261
Armin, Robert, *Quips upon
    Questions*, 1.1.140–4
Armstrong, Alun, p. 2
Arne, Thomas, p. 64
as, 3.2.109, 262
as who, 1.2.45
ashamed, 2.6.35
assume, 2.9.50
assured, 1.3.27–8
attempt, 4.1.417
attended, 5.1.103
attentive, 5.1.70
awe and majesty, 4.1.188
aweary, 1.2.1

backward, 2.2.93
badge, 1.3.107

Index

bagpiper, 1.1.53
bait, v. 3.1.50
baned, 4.1.45
bargain, 3.2.193
bark, 1.1.94
Barton, John, p. 75
base, 2.7.50
bastard hope, 3.5.7
bate, 4.1.71; bated, 3.3.32
bears down, 4.1.211
beast, 1.2.87
beggar, 4.1.436
Bell, T., *Speculation of Usury*,
    1.3.133–4
Benson, Frank, p. 71
beshrew me, 2.6.52
best, 1.2.85; best-conditioned,
    3.2.291; best grace, 3.5.41
Bet-Lessin Theatre, p. 83
betimes, 3.1.19
Betterton, Thomas, pp. 62–3
biblical expressions and allusions,
    pp. 21–3, 35, 40–1, 49 n. 1,
    57; 1.1.75, 94; 1.2.12–15;
    1.3.24, 31–3, 38, 68–85,
    86–7, 95, 157; 2.1.30;
    2.2.62–3, 79–80, 80–1, 88,
    161; 2.5.36, 43; 2.7.69;
    2.9.63; 3.1.80–1, 114;
    3.2.77–9, 246; 3.5.1–2, 17;
    4.1.87, 113, 182, 192–4,
    195–9, 203, 220, 293, 302,
    312–13, 370–3; 5.1.62, 92–3,
    294
bid forth, 2.5.11
big, 2.8.46
Black-Monday, 2.5.25
Blackfriars Theatre, p. 60 n. 1
Blackstone, Sir William,
    Commentaries, 1.3.142
bleared, 3.2.59
bleat, 4.1.73
blessing, 1.3.87; 2.2.80
blest, 2.1.46
blood, 1.2.18; 3.2.176
blunt, 2.7.8
Boccaccio, *Decameron*, p. 14
Boethius, *De institutione musica*,
    p. 23; 5.1.69
bondman's key, 1.3.120
Booth, Edwin, pp. 79–80
Booth, Junius Brutus, pp. 79–80
bootless, 3.3.20

bosom lover, 3.4.17
bottom, 1.1.42
bought, 3.2.311
bound, 1.3.5; 4.1.403; 5.1.135–7
Bowmer, Angus, p. 81
Bracegirdle, Anne, p. 63
braver, 3.4.65
break, v. 1.3.132
breath, 2.9.89
breed, v. 1.3.93
Bridges-Adams, William, p. 71
Burbage, Richard, p. 60
burghers, 1.1.10
but, 3.2.167
by, 1.2.52; 2.9.25; by note,
    3.2.140; by your leave, 2.4.15

Cadiz, 1.1.27
Caesar, Philip, *Discourse against
    Usurers*, 1.3.9, 87
Cameri Theatre, pp. 82–3
candles, 5.1.220
cap'ring, 1.2.59
Carnovsky, Morris, pp. 80–1
carrion, 3.1.33; carrion Death,
    2.7.63
cater-cousins, 2.2.125
Cato, 1.1.166
cerecloth, 2.7.51
ceremony, 5.1.206; ceremoniously,
    5.1.37
certain, 4.1.59
Chamberlain, Lord, pp. 85–6;
    Chamberlain's Men, pp. 28, 58,
    85–6
chance as fair, 3.2.132
chapels, 1.2.13
charge, sb. 4.1.254
Chaucer, Geoffrey, *Legend of Good
    Women*, p. 23; 5.1.1–14;
    Troylus and Creseyde, p. 23;
    5.1.1–14
cheek, 1.3.97
cheer, sb. 3.2.310
cherubins, 5.1.62
choose, 1.2.46; choosing, 3.2.2
circumstance, 1.1.154
civil doctor, 5.1.210
classical references and allusions,
    pp. 34, 48 n. 1, 54; 1.1.7, 50,
    166, 169–72; 1.2.47, 103–4;
    1.3.130; 2.1.5, 32–8;
    2.2.37–40, 58–9; 2.6.5; 2.7.22;

232

Index

235

Index

inderculped, p. 14; 2.7.57
inserted, 1.3.91
inspirations, 1.2.28
inter'gatories, 5.1.298
intermission, 3.2.199
interposer, 3.2.324
Irving, Henry, pp. 10, 68–70, 80
it shall go hard, 3.1.68
iwis, 2.9.67

Jacks, 3.4.77
Jacob, 1.3.68–85
Jacob's staff, 2.5.36
Jaggard, William, p. 90
James I, pp. 59–60, 92
James, Geraldine, p. 79
Janus, 1.1.50
Jason, 3.2.239; 5.1.13–14
jaundice, 1.1.85
Jonson, Ben, *Every Man Out of His Humour*, 1.1.49; 4.1.49
Judas Iscariot, p. 5
jump, 2.9.31

Kean, Charles, pp. 66–7
Kean, Edmund, pp. 65–6, 80
keen, 3.2.274; keenness, 4.1.124
Kemble, John Philip, p. 65
Kemp, Will, p. 60
kept, 3.3.19
key, 2.7.59
Killigrew, Thomas, p. 60
*kind, sb. p. 41; 1.3.138
kind, a. 1.3.173
King's Men, pp. 59, 60
kinsman, 1.1.57
knapped, 3.1.9
know, 4.1.415
Komisarjevsky, Theodore, pp. 72–3
Kortner, Fritz, p. 81
Kyd, Thomas, *Soliman and Perseda*, 2.1.25–6; *Spanish Tragedy*, 2.2.74–5
Kyle, Barry, p. 83

Lambarde, William, *Archaionomia*, p. 24
Langley, Francis, p. 28
last hour of act, 4.1.18
lastly, 2.9.14
lay, 3.5.75
Leah, 3.1.114
leave, v. 5.1.150

left, 5.1.196
legal matters regarding: aliens, 4.1.345; bonds, 1.3.142–8; 3.3.4–5; 4.1.205–7, 232, 244–6, 256, 303; caskets, 2.1.43; deed of gift, 4.1.384–6; 5.1.292; fines, 4.1.350, 368, 377; forfeiture, 3.3.22; 4.1.36; interrogatories, 5.1.298–301; master and servant, 2.2.1–21; oaths, 4.1.225; slavery, 4.1.89; Venetian justice, 3.3.8, 29; 4.1.228; 4.1.37–8, 215–17; wills, 1.2.23–4; 4.1.384–6
less, 3.5.38
let him lack, 4.1.160
level at, 1.2.37
Lewkenor, Lewis, *The Commonwealth and Government of Venice*, p. 26
liberal, 2.2.177; 5.1.226
Lichas, 2.1.32
life and living, 5.1.286
light, sb. 5.1.129
lightest, 3.2.91
like, adv. 1.3.126
line of life, 2.2.154
*lineaments, 3.4.15
livers, 3.2.86
livings, 3.2.156
lodged, 4.1.59
look what, 3.4.51
loose, 4.1.23
Lopez, Roderigo, p. 7
lott'ry of my destiny, 2.1.15
love, sb. 4.1.274, 425; mind of love, 2.8.42
low, 1.3.40; lowly, 2.6.45

machiavel, p. 2
Macklin, Charles, pp. 63–5, 68, 70, 80
Macready, William Charles, p. 66
mad, 5.1.176
magnificoes, 3.2.278
Mahabharata, p. 17
maid, 3.2.198; maiden, 3.2.8
main of waters, 5.1.97
main flood, 4.1.71
make incision, 2.1.6; make shift, 1.2.87
manna, 5.1.294
manners, 2.3.19

The Oxford World's Classics Website

www.worldsclassics.co.uk

- Information about new titles
- Explore the full range of Oxford World's Classics
- Links to other literary sites and the main OUP webpage
- Imaginative competitions, with bookish prizes
- Peruse the Oxford World's Classics Magazine
- Articles by editors
- Extracts from Introductions
- A forum for discussion and feedback on the series
- Special information for teachers and lecturers

www.worldsclassics.co.uk

American Literature

British and Irish Literature

Children's Literature

Classics and Ancient Literature

Colonial Literature

Eastern Literature

European Literature

History

Medieval Literature

Oxford English Drama

Poetry

Philosophy

Politics

Religion

The Oxford Shakespeare

A complete list of Oxford Paperbacks, including Oxford World's Classics, Oxford Shakespeare, Oxford Drama, and Oxford Paperback Reference, is available in the UK from the Academic Division Publicity Department, Oxford University Press, Great Clarendon Street, Oxford OX2 6DP.

In the USA, complete lists are available from the Paperbacks Marketing Manager, Oxford University Press, 198 Madison Avenue, New York, NY 10016.

Oxford Paperbacks are available from all good bookshops. In case of difficulty, customers in the UK can order direct from Oxford University Press Bookshop, Freepost, 116 High Street, Oxford OX1 4BR, enclosing full payment. Please add 10 per cent of published price for postage and packing.